To
Gillian
Welcome home!!

Janis D.

One Way Home

One Way Home

Forrest Dee

A Geneva Book

Carlton Press Corp. ❖ New York

Manufactured in the United States of America
ISBN 0-8062-5108-5

Dedicated to the Memory of

Judy K., Lafayette, Louisiana
Johnnie H., Lafayette, Louisiana
Allen B., Lafayette, Louisiana
Doc Kid McC., Kailua Kona, Hawaii
Frank S., Kailua Kona, Hawaii
and to all who have suffered and died
from the disease of alcoholism and addiction.

Acknowledgments

Special thanks to the following for their very special love and
for being such a big part of my recovery:
Dayton Paishon, Lahaina Maui, Hawaii
Linda Alumbaugh, Baton Rouge, Louisiana
Billy F., Lafayette, Louisiana
Erik Kokko, Honolulu, Hawaii

Contents

Preface

It has been repeatedly suggested by persons who love and care for me a great deal that I have had the kind of life that should be written down in manuscript form, or at the very least I should keep some sort of daily journal. Whenever the suggestion was made, I outwardly showed embarrassment, but my ego ate it up like a kid holding a peanut butter and jelly sandwich, sporting the jam all over his face. Finally, I gave into the temptation and couldn't wait to get to the typewriter to see if I possessed any talent . . . and I think it best for you to be the judge.

I am currently in recovery as an alcoholic and addict. God only knows what else, co-dependent, gambler, sex addict, and skirting on the edge of being an overeater; I really hate to look any further. My sobriety date is October 7, 1985: my rebirth. But for a full effect of the accounting, we have to go all the way back to the beginning. Now it is at this point when listening to AA speakers, my first reaction is to get up and run. I'd like to tell you the definition of an *"ole-timer"* in Alcoholics Anonymous is one who won't go 100 feet to hear someone else's story, but will travel 100 miles to tell his own. I'm afraid I'm slowly becoming an "old-timer." So I thought I would use some bait to entice you to stay and listen. I have experienced three full-blown medical treatments for the disease of alcoholism. Two attempts in halfway houses have failed and several short stays in both hospitals and jails have occurred. There have been brushes with death as well as the loss of many friends and loved ones who have died and are still dying at this very moment, in order for me to stay sober on a daily basis. I have traveled extensively in my life and have made many geographical moves in order to survive, most of which were in vain because I brought myself with me. Today I work vigorously with people in recovery and devote my life to doing God's work. The stories you will read are about people I have met in recovery and how this amazing program of "12 Steps" has effectively and miraculously changed their lives as well as mine. But in order to get to them we must see how I went to hell once, and by way of full circumference, finally found one way home.

People who go to church are people who don't want to go to hell. People who go to AA are people who have already been to hell. I hope you enjoy the stories.

1 Send In The Clown

I want to start by saying that my story and background are by far not the worst. I have known of others who have suffered much, much more intensely than I. As so many of our stories begin, "I came from a dysfunctional family." By this I am not trying to place the blame on anyone for my undoing; only to let you know that my home and parents were having a few problems of their own before I came down to check things out. My mother was an alcoholic and my father was a heavy drinker with a career in the whiskey industry. He used to say that if a man can be a success in the whiskey business he truly would be let into the Kingdom of Heaven. His reasoning was that there was dope, women, late nights, gambling, and of course the booze, so if a person could live through all of these temptations, he would truly be let into the gates of heaven. As kids, my sister and I never had Father home for dinner and we would wait patiently for his arrival and for the fighting to begin when he got home. I can remember as a child that I would have temper tantrums so terrible that in rages of anger, I would beat my head on the floor. Even sometimes as a young child, I would fight my sister as seriously as grown people. My mom said that I would throw baby bottles out of the crib so that I would have her full attention. One of my earliest and most vivid memories brings forth pictures of being deathly sick and needing a blood transfusion to live. Mom and the nurses held me down as I screamed in terror while the doctor applied the needle to a vein in my head, starting the transfusion. I'm not sure how any of this has affected me, but I often wonder if this is where I first began to distrust my mother. The point that I am making is that at this early age I was experiencing traumatic episodes of life and death; involving the human emotion of fear.

I don't remember much about my mom's drinking; perhaps I have blocked it out. What I do recall is being locked in the car for long periods of time while she shopped or paid afternoon visits to other people's homes. Her battles were obviously with my father and they were very physical. Once Dad had to tie her in a chair so that she would stop chasing him around the house with a butcher knife, accusing him of adultery. Doctors were always called, never the police, and the visits to the sanitariums started. In the 60s she went through shock therapy on eight different occasions and medication became part of her life forevermore. She was in a program of recovery, yet was always medicated. This was accepted as sobriety . . . because the pills were prescribed. During this time, I was failing classes and fighting on the school grounds. The tantrums never stopped and the school principal said that he had never seen a child so angry. At the young age of twelve I was sent away to boarding schools. That year, when I went home for Christmas vacation, I visited my mom on Christmas Eve night at the local hospital. She had just undergone another shock treatment and was still out of it. She did not even recognize me. . . . Can you imagine the painful mixture of fear and anger I was going through? What a great Christmas that was for one and all. I can't say much more about that Christmas because to this day it has been conveniently erased from my mind. When I returned to school I picked up my first habit, which was smoking cigarettes.

My father was becoming very successful in business and we were all proud of him. Although his success allowed him to justify his behavior, which took the heat off of him, he was a man of strict rules and frightening ways. Money became his master and he ruled the family with power, prestige, pride, and persecution; mental abuse and abandonment was the norm. He was the Almighty and governed by using material favors to bribe rather than physical means. As my sister received honors for achievements, unreachable for me, I could see plainly that in order to receive the attention I starved for, it was time to send in the clown.

By the time I was a freshman in high school, I had already been out with guys who were all a year or two older than I and had gotten drunk during the summer vacation. There were times when I would pretend to pass out, was drenched with water from a hose at a service station, and laid at the front door of my house. I can still hear my mother screech, "Dick, he's drunk." (That's my dad's name, Dick.) And boy, could she ever screech! For my first six-week report card I brought home five F's and one A in P.E. This was my thanks to my father for allowing me to come home from boarding school to attend public school like normal kids. And that's how I had asked him, too. "Dad, I just want to go to public school like normal kids." Once at a pep rally for the football team, I was called out of the stands holding 700 students to get lickings with a paddle in the back office. Yes, finally I was reaching heights of stardom and would certainly be recognized and soon would be notorious in my little town.

My sister got a brand new 1965 Mustang for her accomplishments at school. Lo and behold, it came with the 289-cubic inch V-8 engine with four speeds and 271 horses. Alas, the magician's rabbit if you will, this baby could get me to stardom in a hurry. I didn't have a driver's license, but that wasn't a problem. I would sneak out at night and steal the car. Not only did I do just that, but I started drag racing the first night. . . . I think I got caught the second night. I found myself in front of a judge for racing cars at extremely high and dangerous speeds *and* didn't even possess a driver's license. The judge couldn't do much to me and didn't want to make a big issue out of it—it might have made him look bad. So off I went. A month or so later, after I had gotten my license, I was out late after the Friday night high school dance. I was burning rubber around a corner near my girlfriend's house and ended up in a twelve-foot ditch. The judge let me off again. He and my dad were pals and I was just a kid getting a poor start learning how to drive. A week later, I was caught laying 150 feet of rubber in front of the local drive-in hang-out. This time the same judge handed me the death sentence; he suspended my license until I was twenty-one years old. It scared me, but I knew I could get my license back. . . . And I had discovered the whole town's attention was on me. "The school was going to the State Football Championship and did you hear about Forrest Dee?" I liked it. Who wouldn't?

The thought occurred to me, more than once, mind you, that if I died achieving this fame, so much the better. That's the price you pay for immortality. I could hear "Remember Forrest Dee?" echoing in my head. I would be dead, but my friends would still be talking about me. My plan was working. I was more famous than the heroes on the football team and I was doing O.K. with the chicks, too. I think they wanted to be part of the excitement. The amazing thing was, that the most beautiful girls in school were going out with me . . . . The rumors were just incredible.

When the girls would break up with me, I would always leave with a bang. Again the ideas of immortality came to mind. I think I really loved some of the girls, if I had any comprehension of the word and its meaning. There was a time I was close to running off and marrying a girl named Marilyn, but her father caught us and ran *me* off for good. I have often wondered if we had eloped, and had a few kids, would I have grown up and settled down? I have learned since that that would not have been the case, but it was fun to dream.

There were several bouts with cars; many ditches, telephone poles, and trees. I remember one time when a friend and I were at a new pizza place that had just opened and *was serving free beer.* We sat with the owner acting like big shots and we left at 11:30 P.M.; both too drunk to drive. Naturally, I took the wheel, being full of grandiosity and ego. I wiped out at a red light like a bowling ball going through pins; three cars were totally destroyed, not to mention the one I was in. I was charged with D.U.I. and wasn't even old enough to drink. By today's laws, I could sue the pants off the bar owner for serving minors and win the case. In those days, I was the only one being sued. My dad was really angry, to say the least, and immediately had me emancipated.

Having gone to an additional military school or two, I found myself back at ye ole public high school in my senior year. I had to retake biology, which was a freshman course . . . and me being a senior, I think the teacher somehow felt that if he could control the senior in front of the freshmen, he would stay in charge of the class. Wrong! I wasn't the guy for the job and was not going to be part of it. I threw my book at him and was promptly kicked out of the class. Not able to graduate with the rest of my classmates, that was the end of high school for me. The senior prom would not be graced with my presence.

Vietnam was looking pretty hot about this time, but I was too scared to join the Marines. Instead I went into the National Guard in order to dodge the bullets. But my father demanded I go into the Navy and somehow arranged it, promising the moon as always. This time if I could produce a military honorable discharge, his kingdom, which he worked so hard to build, would be mine. Off I went; embarking on the fresh start, like the beginning of a new dawn. I graduated with my company as a platoon leader in charge of forty-two men. My parents even came for the graduation. A proud day indeed. Instead of going to southeast Asia, I went to fleet headquarters in London, England. Spinning tales of espionage and importance so believable I started to believe them myself, I spent the next three years there in the heart of England. If we let the truth be known, I was a Navy nighttime janitor, only to graduate to the day shift, guarding the Command Center. With little responsibility and warm British ale, boredom and a freshly broken heart led me to discover dope.

Just as reported in the pamphlets I had made so much fun of, marijuana led to more thrilling and exotic drugs. Speed came next, to be followed by acid, hash, ups, downs, and heroin. Crazier things were used, such as nitrates, poppers, gases, even silver label Right Guard. At some point in all of this, it is hard for me to believe that this is my life I am telling about here, and not someone else's. You see, it always happens to someone else. *They* would do these sordid things and *they* would wind up in insane asylums or in jails or dead—*not me.* Once, after seeing a friend go to the hospital in complete hysteria, I was lying on my bed bumming out from too many hits of LSD. I fully accepted that I would be locked up with the insane for the rest of my life. Later, I was willing to accept death on many occasions from the surge of the needle, from overdoses of pills, or from a racing heart overloaded on crack cocaine beating out of my chest. At a much later stage, I was to reconcile myself to being the village drunk with nothing more to expect out of life. This had to be about the worst thing that could happen to me; not to die and end up being the village idiot.

For a long time the drugs worked and for a while it was even glamorous. I met many so called rock stars and even shared the ritual of getting high with them. While in London I first turned on to some grass that was excellent and I immediately got off. The sound track to the movie, *Easy Rider,* was playing on a cheap stereo, and I experienced music as never before. It was like going from AM radio all the way to a concert at the Royal Albert Hall. Where, by the way, I got to be a frequent visitor to see many rock and roll bands. I had a small flat in the Chelsea area, which was a cool place to be in London, and would see people like Rod Stewart, The Rolling Stones, or Joe Cocker, on occasion. I had stars dropping in on parties. At one late night party I saved one from jumping out of a window, who was celebrating his Oscar, high on acid.

By the time I did leave London, all of my friends gave me a poster of the poem by Rudyard Kipling entitled "If." Then immediately presented me with a full-size Russian commie flag. I didn't know if I was supposed to go back to the mainland U.S. and start a revolution or what.

It was great to be part of one thing or the other, whether Woodstock, the Isle of Wight, or the protest culture itself. I couldn't wait to get out of the service, grow my hair long, and tell the whole world, "Damn you world, you've got it all wrong." The Navy let me out early after admitting me for drug treatment in San Diego, California, and making me promise never to come back! I give them credit. They really did try to clean us boys up before they sent us home. Even though in the experimental stages, this qualified as my first drug treatment. *Most of us had no intention of getting clean.* It does definitely qualify as the beginning of some of the nut houses that would be coming later into my life.

I did indeed bring home the honorable discharge, but abracadabra, the kingdom wasn't mine! Disappointment and disgust were all over my family's faces. To top things off, in another thirty days, I was sick with serum hepatitis and was down for the count. Totally dependent on my mom to feed and care for me, I started a sentence of sixty days of being bedridden. But after forty-two days, I couldn't stand it anymore. I had a hippie girlfriend in New York City who had vowed to take care of me. Against doctor's orders and with threats of death, I packed and left for the Big Apple. After being there for three whole days, this woman of outstanding character left me and to this day I don't know how I survived. Going into supermarkets and coming out empty-handed seemed to be the order of the day. Weak and yellow, the struggle for health went on.

New York City got to be reasonably comfortable after I regained some color in my face. I was drawing unemployment like every other veteran who had just been discharged. I was fortunate enough to meet some of the more famous unemployed actors and actresses. I went around pretending to be a musician and since I knew a few chords and had some equipment, they bought it. The problem was, so did I. I wasted the next five years of my life trying to play. In reality, I was doing quite well getting loaded and dealing drugs, thank you. One day a package of dope didn't arrive and I knew the bust was coming. I had been in the city for a year and had seen enough to know it was time to go. I shipped everything out and caught a quick plane home. After arriving on Tuesday, to a family that seemed accepting, I was busted on Wednesday by Federal Agents. Welcome home, boy!

Never to do any time for the crime and always getting off through one loophole or another, school looked to be the next comfortable alternative. Using the G.I. Bill, I managed to flunk out in record time. Again I blew my dad's kingdom by not bringing home the good ole college degree. The bright spot was meeting my first wife, getting married and moving to Austin, Texas. The dope business was booming there and I had good interstate connections. Here we go again! Pot runs going to the east, with mushroom runs going to the west, serving a few college towns like Norman, Oklahoma, to the north were normal. I embarked on a new and illustrious career of smuggling. Packing guns and riding in pickup trucks came to be the status symbols of the West for me. The Feds had me once on the border coming in at McAllen, Texas. They took the grass, left over from a sample, and told me never to be seen down there again. There must have been ten or twelve Texas Rangers and my buddy Fernando and I were happy to get out of there alive. After all, the ton of grass had gone through in a dump truck before us.

My dad experienced a heart attack and was requesting my presence, so I made the visit home. Once more, he held the carrot in front of me. It would be foolish to allow his partners in New Orleans to gain control of the business if he were to die. It seems that these men had cut an agreement. If no living heir was actively involved in the business at the time of death, the business would be sold to the living partners for face value. I packed up my life and moved to New Orleans. Can you see a person with my history becoming a whiskey salesman in New Orleans? It almost killed me and after a year of that, my life was in shambles. Separated from my wife, fired from the job, broke, strung out, I spent a short two weeks in treatment.

While in Turo Infirmary for a two-week evaluation, we had groups and one-on-ones with the doctors. I was in such a fog, I can't tell you how they released me with a clean bill of health. I remember meeting a nurse who lived in the same apartment complex as myself, and we started to talk and get along grand. Soon enough, this nurse was fixing me with shots while I was in treatment and when I got out, it continued at her apartment. Once we rigged up a cocaine solution to an I.V. system. I sat there watching the drops going into my arm until I went so crazy that I started breaking everything in the apartment. A shot of Vestivol in the cheek of the ass was necessary to calm me down.

Now divorced from my first wife, I fell in love with the manager of my apartment complex, moved back to Lafayette, my hometown, and went to work in the oil field. I married this beautiful woman, she had a baby girl with me, and by some total inability to love anyone, I destroyed everything. Skipping the details, I will share the worst night of my life with you. The day my daughter was born, I just didn't know how to handle it. . . . I spent the entire night with a hooker. After leaving, at 4:00 A.M, I challenged God, "If there is a God, take my life or spare it, but do show yourself and show yourself now." I jumped a high ditch at over 100 mph, that Evel Knievel would have been proud of, only to have God spare my life once again. The car landed on the other side of the ditch, plowing down pine trees and breaking the car in half. I emerged without a scratch and without atheism, that's for sure.

The oil field didn't work for me, although I did spend a lot of time attempting some sort of stability. I finally quit work, going into dealing dope full time in the cocaine business. There were huge amounts of cocaine, but most of it was used for ourselves. Every once in a while I would go on runs that would last weeks. One particular partner of mine could go on a run forever. We would both arrive home looking like walking death. Once his wife called me, saying that Billy had locked himself in a closet and wouldn't come out. He had a big closet with a window in it and I knew exactly what he was up to. She expressed grave concern, asking if I would come over and help Billy. I consoled her and without hesitation rushed to the rescue. Once I got there I ran upstairs, knocked on the closet door and called out, "Billy." He asked, "Who is it?" with a mean, frightening voice. "It's me, Forrest," I replied. Again harshly, he said, "What do you want?"

"Billy, I don't want you to come out. I want you to let me in." He did, and off we went on a tout that lasted for days. The next time his wife saw me, I think she could have killed me.

I want to tell a few more war stories just to show you what kind of drunk I really was. (Lately at the meetings I have attended, it has bothered me that there are so many people in attendance, coming from what is known as a "high bottom." High bottom can be defined as those displaying alcoholic behavior and / or casual drinkers out of control, therefore causing an unmanageable lifestyle. I really don't think that they should go out and drink anymore, but AA is being used as a social club instead of a life ring for people in dire straits. I certainly would not have quit drinking had my parents shown up with the family pastor on a Sunday morning.) These stories will help you understand why I hold AA so dear and care so intensely about our program.

The second step states: "We came to believe that a power greater than ourselves could restore us to sanity." Most drunks that I meet have a problem recognizing the insanity in their behavior, and I certainly have suffered from the same denial. These stories confirm how desperately I needed steps 1 & 2. In 1977, I was in New Orleans and had been drinking all Friday afternoon. The joint was downtown and it was getting to be evening; time to go home. I walked out, started the car and immediately was in bumper-to-bumper traffic with miles and miles ahead of me to get home. I pulled the car onto the right side of the highway and started down the shoulder picking up speed, flashing emergency lights, and shooting my .357 magnum pistol out of the driver's side window.

Later in my drinking career in Houston, Texas, a car cut me off on the freeway about 10:30 on Friday night. I jammed him against the concrete barrier separating the eight lanes of traffic and squeezed him against the wall at seventy miles an hour, leaving sparks to light up the median for a few 100 yards. I pulled away leaving him in a wreck. Someone immediately started to tail me at a high rate of speed and since no blue lights were flashing, I figured it was some Texas flathead wanting a piece of me. I was sure willing to give him as much as he thought he could handle. I caught the next exit and watched him follow the lure. When he came down the ramp behind me, I knew I had him dead in his tracks; he couldn't go backwards. Hitting the brakes, the truck did a doughnut and left the vehicles facing each other. I grabbed my pistol, opened the door and put a bead on the first guy getting out of the car. He had a piece drawn as did his partner and he shouted, "Police, drop it! Put your hands against the truck and spread 'em." They weren't in uniform, but it was obvious they were Houston detectives. They grabbed my license and ran it through the computer, called for back up and directed me not to make a move. Somehow they didn't cuff me; their first mistake. Their second mistake was when they gave me back my driver's license. At their first lapse of attention, I hit the cop as hard as I could and was back in my truck, burning rubber down the road making my getaway! I wasn't going to let 'em catch me, no way! It would have made the Rodney King incident look like a tea party. They would have killed me; I'm sure of it.

Sometime later, Billy and I went to New York on a trip to collect funds that were well over six figures. We got there and made immediate connections for the best quality cocaine in the continental

United States for us to use while there. We went to some hideaway motel in Brooklyn. We did *lines* for a few hours. I cleaned up and left Billy, anxious to get to the night life in Manhattan. The night passed by like a fogged dream and after several drinks and a couple of hookers, I arrived back at the room at two o'clock in the morning. I opened the door to find Billy hiding behind the curtains with a loaded revolver in his hand, ready to blow up anyone who might have followed me up to the room. It didn't take long for me to figure out he had been in the powder room all night. After a couple more hours of packing his nose, he started to panic and looked to be in trouble . . . calling out for a doctor. I told him, "Billy, don't make me call a doctor to some Brooklyn, New York, motel at four o'clock in the morning. It will cost a lot of money and use up some favors I don't want to use." He looked like he was going to die, so we made the call. I rushed him in the back of a cab down to some doctor's office and he met us there. "Doc" shot him up with a syringe of Thorazine in each cheek of the ass. We were comfortable again; long enough to finish our business.

On the way home I became very resentful about the entire weekend. After all, I was a front line lieutenant and not a babysitter. Billy had the audacity to accuse me of doing too much cocaine. By the time the plane landed in New Orleans, I had already figured a few tricks to pull out of the bag for myself. That night I got home to my wife and we stayed high all night. We packed our bags and chartered a jet to Vail, Colorado. The next morning I was going to be sick coming off of the drugs and I was either going to need a resort with servants or a hospital. I took the resort. I passed out when the plane reached a very high altitude. The pilot put an oxygen mask on my face and had an ambulance standing by for our landing. I woke up asking what all the commotion was about and they told me I had turned blue and quit breathing. I laughed it off, paid the pilot in cash and started to bitch about my limousine being late. Needless to say, it wasn't much of a fun trip; I barely made it out of the condo for any activity at all. When we got home, Billy was fuming and looking for the $10,000 missing from our funds. I blew him off as if to say, "That's the cost for babysitters and telephone calls for doctors at four o'clock in the morning in New York City." In reality, I felt the trip was a bonus for putting my ass on the line. Another partner, named Mike, was doing the accounting one New Year's Day and said I had gone through a quarter of a million dollars in coke in the previous year. He wasn't happy, but I damned sure was.

Once I set out to kill a man named Charlie for ripping me off of a couple ounces of coke. I had tracked him down to Gadsden, Alabama. I brought along Sammy, whom I was involved with at the time in the dope business, and it's a good thing he came with me because he stopped me from making the biggest mistake of my life. We broke into this Charlie's parents' house and caught him trying to run out the back door. After a wrestling match, I knelt him down and stuck a large caliber colt down his mouth, cocked the hammer and was on the squeeze when Sammy pushed me off balance. For a split second of sanity, I hesitated, looked down at this bag of shit at the end of my gun, and eased the hammer to its firing pin. I let him go, bought some whiskey, and drank all the way home to New Orleans. Elvis had died that day and I wasn't in the mood for any pain. By the time we hit the causeway at Slidell, going westbound into the city, a load of blacks pulled a pistol and pointed it at our car. *Heeello. Big mistake!!!* I pulled off the road while they were all laughing like they had scared a couple of white boys. Pulling my .30 caliber carbine with thirty rounds in the clip, I let Sammy drive and I sat shotgun. He pulled the cars even and when I stuck my head and the carbine out the window of that black Pontiac Grand Prix, there were six blacks who had all turned a pale shade of white. The bullets started to pepper the concrete guard rail that we had them pinned against, then covered the hood, and glass started to fly. I aimed just high enough not to hit any of them. Who knows? . . . . I didn't stick around to find out, nor did I give a damn. We pulled into the "Big Easy" that night. I didn't mention the incident to my wife nor did she know where I had been. I watched TV, "Mourning the Death of the King," and for me it was just another day at the office. Later I was to lose that same carbine to the Lafayette Police Department. I went crazy one night and leveled a "drinking hole" with the rifle after some guy had hit me from behind and opened up my chin to the tune of twelve stitches. Not so very far away from that incident, I was kicked out of another apartment. I was playing with one of my pistols and it misfired and shot through the next three apartments. The manager asked me how I would have felt had I killed someone and my reply was, "I guess I would just have lived with it." Insanity at this point was rampant. I was watching myself on the news and reading about things in the papers that I knew more about than the reporters or the cops. I knew what the D.E.A. was going to do before they did it . . . it was getting to be a deadly game now.

We rented entire hotel floors for the dealer conventions that were springing up in Atlanta, Houston, and New Orleans. At four o'clock one morning, I was sitting in the penthouse of a swanky New Orleans hotel with a *key of powder.* One of the bosses was crying on my shoulder about how lonely it was at the top. It just didn't feel like the top to me. Somehow I could sense the beginning of the end.

Things were getting close to the end. It was the late 70s and freebase cocaine was now on the scene, changing people I had known do cocaine for years into instant junkies. Even Billy fell to the addiction of the pipe, ending up in treatment himself. My second wife, Sally, took our baby and left for good. Who could blame her? There were moments of insanity when I shot up our home and was arrested for wife beating. Seventy-two-hour evaluations were common, to make sure I was not a danger to myself or others. She was smart to get out with her life and save the child. It didn't take long; maybe thirty or forty days and I was up holding a gun to my head . . . but I stopped long enough to make one last desperate call. The call was to my mom, and she, in turn, called Billy. Off I went to treatment in Baton Rouge, Louisiana. It was time for the clown to die. I'm not sure exactly when he did die, but he is dead now, hopefully never to reappear.

2 Then Came Hell

As they rushed me off to treatment, I was feeling scared, desperately lost, hopeless, and a little nuts. Funny enough, I had been through some of this before, yet I didn't know what to expect. Billy was driving a Mercedes at the time and I was trying to enjoy being seen in the car, not thinking about going to the nuthouse. When we arrived, they seemed to have a problem finding room for me! Was I to be accepted or not? *Can you imagine that?* I mean, don't these people know who I am? If this facility has no beds because of their lack of foresight, I *could* go home. I was okay, now. Chalk it up to a weak moment of the booze with gun in hand . . . it wouldn't happen again. Then somehow I found myself in an office answering admission-type questions. They must have put some other poor soul on a cot in the utility room to make room for me . . . a real addict, a special case.

I spent the night in reasonable conditions, sharing a semiprivate room with a fellow who looked pretty clean and somewhat hip. It would do. The next morning there was some sort of prayer followed by kindergarten calisthenics, and then breakfast. I was checked out by the doctor who gave me a clean bill of health, with the exception of nasal scar tissue which might require surgery later. Wasn't it great to have a doctor confirm what I had been telling people for years. The b-i-g cocaine user . . . what an ego. And later I found out it had been killing me and I wasn't through yet; the nightmare was only beginning.

Next it was time for "group." Ah, yes. I've done this before . . . just watch me strut. This counselor has never seen the likes of me before. Should I put them in full shock right away or just be cool; giving them that insane look? As if to say, "one wrong move and you're dead." In walks the teacher, and guess what? I get a substitute. I was beginning to know how the guy felt that walked in to buy a car and walked off the showroom floor with the only lemon to come out of Detroit that year. . . . Or, the poor slob that stands in line for hours to buy tickets to the show and he finally gets up to the window just as the "sold out" sign goes up . . . I bet if there was one pile of dog shit in the yard, I'd be the one to step in it.

Well, that's about right. I was dying and they send in a *substitute.* That's it; I just dare them. I'm going for the throat on this one. Ready to explode, I sat squirming in my chair listening to a lady crying about something as petty as thinking she might be an alcoholic. Then some guy talked about his relationship with a girlfriend going on the rocks. Another one was concerned about losing his business because *possibly* he drank too much. Smoke must have been coming out of my ears and I could feel the halo of fire around my head, like what a shot of heroin gives you after you've just boosted. When the counselor, Linda, turned to me and sarcastically asked, "What *is* your problem?" She had no idea of the wrath about to be unleashed upon her.

She didn't deserve to know what my problem was, nor was she qualified to solve it. She might be able to help these amateurs figure things out, but I could see I needed special consideration. With as much disgust as I could muster, I demanded to know her credentials. "How on earth can you help me? You're not from my world. Did your books tell you what it's like to lose your dignity, family, love, everything in the universe? Tell me what it's like when after losing all of your life, the one thing you want more than anything is a fix. You explain to me how you came from *there* . . . to here, well-adjusted, family intact, educated and qualified to help me. Then I'll listen. But if you haven't come from the same hole I am in now, we've got nothing to talk about."

She looked at me and there was no anger in her face. She began by telling me that she was the head of this treatment facility, standing in for the regular counselor who was on vacation. Yes, she *had* gotten most of her knowledge from books, but had been successful in helping a lot of drunks. She wheeled her chair over to me, putting her arm around me and said, "I know the pain you are in." She then shared a small part of the pain she had experienced in her childhood, discussing insanity in her mom and abuse from her dad. She held me as I broke down into little pieces and started to cry. I realized for the first time that alcoholics aren't the only people out there who suffer. Some people have very serious problems to overcome and actually do it without a bottle between their legs. The group supported me while the counselor and I started a spiritual journey that would become a lifelong relationship.

I gave what I thought to be the most honest and enthusiastic effort I possibly could at this hospital. For the next thirty days, we were given a vivid education on what we were afflicted with. The disease of alcoholism, over which we had no control. There were group sessions on a daily basis and one-to-one counselor-to-patient relationships. We were tested in numerous ways to enlighten the staff about our personality dysfunctions. Linda had called me into the office to go over my test results and I remember being asked if I believed myself to be a paranoid person? My response was, "Not really." She looked at me with astonishment and said, "You went off the scale." She used as ammunition, one of my stories that I had shared in group, against me. She said that anyone who has driven, through his rearview mirror, from Houston to New Orleans, goes off the scale in the area of paranoia. She was right. I was all messed up and couldn't even see it.

They found all sorts of things wrong with me; from learning disabilities that had affected me as a child, to my delusions of grandiosity. Brain damage was a concern, but somehow or other I didn't have any. Funny, I still think I have plenty of that, if nothing else. Soon it seemed I was on my feet with a rather good footing. I felt that I was at the head of the class for maybe the first time, at anything, in my life. I just knew that I was to pass the course with flying colors, but it wasn't to happen. Matter of fact they didn't even graduate me. Linda started looking for places to send me; and would I consider going to a halfway house? She dropped the subject and I took it as a test.... Would I go to any length?

Linda had smashed any ideas I had about getting my wife back; somehow she knew that it was a done deal. I, for the life of me, couldn't figure out how she could make such a harsh judgment on a surface level. In two weeks, when I was allowed to use the phone, I made the call. I found out what Linda already knew. The pain can't be described. Breaking me down to my knees, I selfishly petitioned the Lord in prayer for the only thing I knew that would take the pain away other than alcohol. I asked for another woman. God had shown himself to me before; while in screaming fits of rage, dancing with death, I had commanded his presence. Surely this small task was a mere trifle for him to perform.

I asked and he gave. The next day, in walked Jamie, the woman I would spend the next twelve years of my life with. Years later, I heard that the second dumbest thing in the world was for two newcomers to fall in love. And that the dumbest thing was for their sponsor to try to stop them. Well, they said our chances were a million to one. Naturally, I said I liked those odds. Jamie and I put things on hold and I got through this treatment and was sent off to a halfway house in St. Paul, Minnesota.

This particular place looked like a criminal house for waywards and actually, with a closer look, that's exactly what it was. Can you imagine that the treatment center had this image of me? There was some "long hair" running the place that was more grandiose than I was. I think he believed that God had died and left him king. Who cares! I'm out of there and on my way home in two weeks. I called Linda to let her know things did not go well and she arranged for another place, this time in Omaha, Nebraska. Off we go into the wild blue yonder all over again. Hopefully this place would be reasonable, and it was. This time I lasted about sixty days and blew out when another guy from Louisiana and I decided to hit the bars in Omaha on a Saturday night. We were walking down the street with no intentions of misconduct. Passing by a place where the band could be heard on the street, we looked up and saw a sign that said, "Free Beer Tonight!" We waltzed right in. It was incredible how fast it happened, and within hours we were drunk and smoking a joint.

Taking one look at me loaded that night, this guy said, "Wow! I can see why you get stoned." I seemed to glow with power. Like a star before he hit the stage, I was *on*, again. To hell with this straight stuff! It's not for me! They asked me to stay on at the halfway house, but you know when you're not going to make it. So why bother?

When back in Louisiana, I hit town low-keyed. Trying to hide any evidence of my slips, I drove to Baton Rouge every chance I had to be with Jamie. We finally moved in together and made an attempt to go straight that lasted about a week. That weekend it was . . . Jamie and Johnny Walker Red and me. Our first fight over money matters pointed me back in the direction of Lafayette, and I made many attempts to recapture her fancy. Jamie would see me between dates with other men, driving me crazy with jealousy for years to come. The beatings began and became progressively worse. How we didn't kill each other must have been due to God's intervention. The human mind can play funny tricks while obsessed with jealousy. At times while waiting for sleep, I would spend hours plotting how to catch Jamie in the act, kill her *and* get away with it.

Strangely, I believe I really loved this woman. *They* say that sick attracts sick and they are definitely right. Besides having the same preference in whiskey, we shared much in common. She competed fiercely and so did I. We both enjoyed driving in the fast lane, dangerous sports, and going to happenings; she worked hard and had some pride left. She was to never get over her resentments toward me even after years of sobriety and my attempts to make amends. She never did let go of those horrible nights of rage between us nor the suffering that lay deep in our souls. Later it all turned against us.

There were jobs here and there, but nothing lasted very long for me. Finally I landed a good job with an international tool company. There was a company truck and a little Creole style house on the outskirts of town, just for Jamie and me. Vowing to go on the wagon once more, we intended to have a life of normalcy. Why not us? Why not . . . just this once.

After 120 days, the company gave me the weekend off and my buddy and I headed for the Gulf of Mexico on a fishing trip. My dad had a camp at this resort community, but we weren't going to bother him for it. We would rough it. We stayed up most of the night drinking and getting loaded on pills. After having made most of the last 120 days straight, I was handling myself pretty well and was in control. I have already shared with you the most horrible night in my life. Now, you are going to hear about the day I regret the most. Leaving the camp at the crack of dawn, we set out for a day of fishing in the Gulf. The weather looked good, the seas were calm, and I started to enjoy the first moments of a long weekend away from the pressures of the job. We weren't two miles out when the outboard motor stopped abruptly, leaving us dead in the water. We finally got a tow after drifting for half the morning. I guess it was about one o'clock by the time we arrived back at shore. I was pretty calm through the haze of a dozen or more beers, when I spotted my dad's car at his fishing camp. I thought I could still salvage the weekend by mending ways with my dad and in the meantime, I could talk him into letting us take his boat out the next morning. Needless to say, he wasn't very happy to find me knocking on the door with a beer in my hand. He literally turned me away; asking that I leave and stay out of his life for good. I told him I would stay out of his life, all right . . . after I showed him exactly what I thought of him. You can figure out the rest. And I spent the night in jail.

Jamie bailed me out the next morning and we spent all day getting drunk. If beating up my father and landing in jail wasn't enough, I put the company truck in the ditch, rolling it one and a half times and landing it on its top. After maneuvering a trip to the hospital instead of receiving charges for D.U.I., I managed to sleep in my own bed that night. Monday morning came with the phone ringing and it was my boss with bad news. The oil business was slowing down and he was going to have to lay me off. He ended the conversation by telling me to bring in the company truck. Following his bad news, I shared some of my own —he could go and get his truck that had just been totalled, from the junk yard. He was not happy to say the least.

With rent due, plus electric, gas and phone bills waiting to be paid, I needed to make a quick score as the whole world was fixing to cave in. I knew a guy who would still front me some dope. I picked up 500 quaaludes and went to Baton Rouge; selling off 200 of them that night and winding up in jail on a D.U.I. By the week's end, Jamie and I were killing each other, as usual. She took off that Saturday evening and as the sun went down, I hoped I'd seen it for the last time. By 7:00 P.M. I had eaten thirty or so quaaludes, chugged a bottle of Johnny Red, and cut my wrist with a hunting knife. I opened the biggest gash in my arm you ever saw and it hurt like hell. Jamie came in about 10 o'clock to find me unconscious in a pool of blood. I had left a suicide note to the little girl who had been taken away from me; leaving her all my worldly possessions. All my worldly possessions, what a laugh! The phone and lights had been turned off and I was soon to be kicked out of my rental and had no job or worldly possessions to leave. I stayed unconscious for three days, waking up with tubes down my nose and throat. I was black and blue from the restraints tying my arms and legs to the bed. My first thought was, "Oh, shit! I lived!" My second thought was, "I'm in big trouble now. They are going to lock me up in the nuthouse and throw away the key."

They stabilized me so I could be sent to a holding facility, then to treatment at a famous institute. One of the doctors at the temporary site saw me and started to shake his head in dismay. I pleaded to him, "Why can't I get it, Doc? Why can't I get sober?" He looked at me and said, "You're hard core, Forrest. One of the worst cases I've ever seen." There was no doubt I should have already been dead. Before I flew to a treatment center in Kansas, my dad came to tell me that he was going to spend the money for the absolute best treatment for me. If that didn't work, he was at least going to be able to go to his grave knowing he had done everything earthly possible. A monsignor from the parish came to see me off. He talked with me and told me the story of the Prodigal Son. Blessing me and the journey ahead, he called me the Prodigal Son and left me standing on the runway. I really felt special and maybe I was. No one could have defied death as I had and been spared over and over for no good reason. Away I went on a private jet to the nuthouse. Just like the ride in the Mercedes with Billy, I was more intrigued by the equipment than where I was going.

The treatment lasted six weeks and did not go well. The doctor in charge made a brilliant statement the day before I was discharged. He said, "If you continue with this lifestyle, you will die." Great! I had entered the place after a suicide attempt. I figured any second grader could have come up with a better prognosis than that. The family counselor, whose job it was to mend family bridges, hadn't even made a phone call contacting any of my relatives. Two weeks after I had left, this particular counselor was in treatment himself for alcoholism. I began to build up an incredible resentment toward recovery centers. After all their hocus-pocus recovery, they tell you to go to AA anyway. I could have done that from the start. I was very angry and wanted to sue them all for malpractice.

Two events happened while I was in treatment that had nothing to do with treatment at all, but struck me tragically hard. The first was a call from Jamie. She had gone to visit Judy, a friend of ours. I had known Judy since high school and we had all been trying to get her straight for a few years. We would go to a meeting and smoke a joint before and then after. She had seen me go through hell and I had seen her there also. I guess you could say we had a mutual admiration for each other. Jamie was telling me that Judy was in an accident. There was a hesitation in her voice and I knew something was seriously wrong. Jamie said she was dead. It seemed that she had scored some tuinals from a notorious local doctor and had fallen in the path of an oncoming car. I was so upset that the hospital had to give me a sedative.

The second event came when my dad arrived by plane to get me. We spent the night at a local hotel and we left for home the next morning. Before going to a restaurant for dinner that evening, he sat me down to tell me that a good friend of mine named Allen had shot himself in the head and was dead. The news left me in shock. I never seem to know how to react to these things; it is painfully heartbreaking. A feeling of hopelessness came over me . . . I felt none of us were going to get out of this alive.

Returning home, I took jobs from shucking oysters in restaurants to being a newspaper agent in charge of kids with paper routes. I figured what's the use and lasted about 120 days sober after release from treatment. I remember coming home to Jamie one night totally bent out of shape. "What's wrong baby?" she asked. I told her that I wanted to drink so bad that I was ready to jump out of my skin. To my surprise, she sat me down on the couch and went to the kitchen and *mixed me a Johnny Red.* Seems she had just picked up a bottle that night. Instead of being mad, I could have kissed her.

After another month or two, my dad moved me to a 220-acre country home that he had vested interest in. I was to care for the place; cut and hedge the yard, mend fences and look out for vandals. I think the truth was to just keep me out of the public eye; till he figured out what to do with me.

After another horrendous fight with Jamie, she moved back to Baton Rouge. Her father and brother helped her pack and both of them had come carrying pistols . . .I didn't blame them. Soon the loneliness crept in like never before. I made thirty days clean, back out, ten days, back out, two weeks, back out, on like that. Whiskey sores were starting to appear on my body. I walked into a meeting one night with a huge weeping sore right between my eyes and an old-timer said, "Forrest, if I had sores like that I'd quit drinking too." I heard things at the meetings like, "You just don't have to drink again." And even that went over the top of my head. I just couldn't fathom it. My dad came to see me shortly after reading my name in the newspaper for shoplifting a rib eye steak at the grocery store. He wanted to know what was I going to do with my life, as if I had a choice. "I'll get a job," I said. He mentioned that no one would hire me so why didn't I get away; as far away from the family as I could get without using a passport and suggesting Hawaii, he then produced a one-way ticket. He was tired of me embarrassing the family.

I took him up on the trip. After all, I was going no place fast. He looked at me with tears rolling off his face, saying, "If I were you I'd get a fresh life, a new start, a new lease." I smiled as if to reassure him that's just what I would do. Get a fresh start, go off and become somebody and with a big hug, we parted. Two weeks later with my entire life packed into two suitcases, a jet had my attention again. Not knowing anyone or anything about where I was going, fear and pain were my master. I got off the plane so drunk I could barely walk. It was late when I caught a cab to the Waikiki Holiday Inn. After checking in, I went out trying to get laid for a big aloha welcome to Hawaii.

Out 'til 4:00 A.M., so drunk that the hookers didn't even want me, finally I called it a night. When I woke up, the first thing I did was to jump into the Pacific Ocean for a swim. I was so excited I jumped in with half of my clothes on, losing a new pair of Ray Ban glasses in the surf. Fifteen minutes later I was scoring some of the best pot in the world. I bought a newspaper and went to my hotel room to search for apartments. Firing up a joint, I literally got so high off the wacky weed I couldn't read. There were so many Hawaiian words in the paper that denoted areas and subdivisions, I literally could not read the paper. After a map, another joint, quick inquiries and directions, I was no better off. The one thing that I could read was how much *money* it would cost for rent. Immediately, I knew that my budget wasn't going to last long at these prices. The Holiday Inn could have me broke within the first week. Somehow I managed to pick up the phone and call a realtor, and within forty-eight hours I had a place that was fair, but nothing spectacular. I paid $1400 for the first and last months' rent and I was in. After the first week in Hawaii, I was broke and had to take a job cleaning floors at night. Within another two weeks I was selling cocaine on the beach and hanging out at the Harbor Pub for entertainment. After almost two months in Hawaii, I had not even been around the island yet; my whole life consisted of the apartment and Harbor Pub.

One night a fight broke out between me and another guy over a dope deal. I found myself holding him over the edge of a twenty-eight-story balcony and damned near threw him off. I woke up the next morning realizing that I had brought to Hawaii the one thing detrimental to myself and couldn't shake it . . . I had brought Forrest. I found myself frantically banging quarters into a pay phone calling AA. A guy picked me up that night and took me to a meeting, right across the street, the Waikiki Monday Night Men's Stag Group. I had arrived and it was there that I first felt I belonged.

I would like to report that I was there for good, but this wasn't to be the case. There was even a heavier price yet to pay for my sobriety. The good news is that I joined the group and did enjoy some recovery for the first time. Moments of loneliness came and at these weak times I would always pray for yet another relationship. I got a great sponsor and seemed to be on my way. My interest in boats landed me a job in the fishing fleet, and I would spend two to three weeks per hitch at sea. Money was good and like my old self, I spent it quickly when I hit the beach. This time I didn't support any drug habits for myself, but I did manage to start getting back some of the things that I had thrown away in my past.

One Christmas I had given Jamie a Rolex watch identical to the one I wore. I presented her with the gift and she had nothing in return for me. Her mother was so embarrassed that she ran upstairs to wrap a shirt that she had planned to give one of her boys and presented it to me. After that Christmas, when Jamie and I had a fight the watches would start to fly. When she and I had finally broken up, I took my watch off my wrist and threw it in a ditch. So when I started rebuilding my life in sobriety, the first thing I bought myself was a new Rolex watch. I still have it today; a special gift to myself that I worked very hard for. But if you look deeper, the watch really covers a scar; a tattoo that reminds me how desperate I once felt. I have been embarrassed at times when the scar on my wrist shows, even by the simplest task, such as reaching for change from a grocery cashier. I have always felt pretty selfish and cowardly for wanting to check out on life. So I covered the scar with the best timepiece made by man to remind me that I am on borrowed time, that life is very fragile, and that time is precious. People who love me see the watch and know without any question what it's about. Others react with envy, some with sarcasm, and some with respect. Those people who react out of respect, look into my eyes and know that I have earned it. The symbolism in my life is deep-seated; the car I drive, the jewelry I wear, etc. The environment I live in is extremely important to me. So the next move I made in recovery was to change where I lived. It took most of my paycheck, but I wanted a nice place. Especially a home where I had not tried to kill a man.

My new condo would not be ready for a month and I had already terminated my lease at the old one. So one of the guys from the group offered me a place to stay for a month, or however long it took. He had about a year of sobriety, with tales of receiving last rights on his deathbed to back his story up. His name was Robert and I guess he was about fifty-five years old and French Canadian. He knew that under no circumstances could he ever drink again . . . for him, it was to die. Yet, while I was out to sea, Robert got into the vodka and died within two weeks. It affected me, but this time I did not grieve. Had I turned this cold? Like a Vietnam soldier, not getting too close to buddies because they might turn up, and did turn up dead? Apparently this is exactly what was happening, and it would get worse before it got better.

I moved into my new home, a luxury Condo on the forty-third floor. The view was incredible and I was high enough that the screams of a Waikiki Saturday night would not reach me. I could see forty miles away to the next island. The colors were wonderful. The newness wore off in a couple of weeks and the loneliness set in again. I was still obsessed with having Jamie back in my life, sharing with her what I knew in sobriety was going to be an exciting life of miracles. I called her and found that she had gone in for another treatment herself and was currently back making meetings for about thirty to forty days. I just *knew* that we could make it and everything would be all right.

There were some dates between calls to Jamie, and I even met some nice girls in the building. I worked hard so I could send Jamie a ticket to meet me for a week's vacation. We had talked about it and the ticket was finally in the mail. She would come and share paradise with me at last. Three days before she was to arrive, excitedly I called to say "can't wait to see you." There was silence . . . Oh boy! Here we go again. "Okay, Jamie, what is wrong this time?" As I held on tight, she explained that over a cup of coffee at an AA meeting, she had met someone new. I hung up the phone, hurting so bad I doubt I'll ever forget it. I prayed and asked Jesus to allow his Mother to hold me just for a moment until the pain would go away. I was crying like a baby, and what's more, I hated AA and everybody in it.

Not long after this, God sent an angel to me. She was a queen—good looking, caring, and had a daughter the same age as the one that had been taken out of my life. Wendy was perfect. The problem was . . . I wasn't. I owe her amends today for not treating her right. I was eight months into recovery and things seemed to be going really well for me. I had rebounded financially and even mended some bridges on the home front. My dad was so proud of me he financed my very own fishing boat. And I was in business.

With money left over, I bought Billy a ticket to come see me and made another attempt to get Jamie to come over. It seemed that the AA Club House relationship of hers hadn't worked out and she could use a vacation. Billy came over and we did some fishing. He met Wendy and approved. Yet things didn't feel quite right. By the time he left, I was drunk again. Maybe I had gotten too much, too fast. Maybe I was still in pain from Jamie. Maybe Wendy had to go. Maybe sobriety was boring and I just wanted some action. Maybe . . . Maybe.

Jamie did come for a week and things worked out well. Naturally, they worked out well. I had my drinking buddy back. After Jamie left, I broke up with Wendy and made long distance plans for Jamie to move over, permanently. I would go home for Christmas and bring her back to Hawaii. Jamie had just lost custody of her kid and she was ready to escape off the edge of the world. That holiday season was horrible. I was hiding my drinking from everybody, just waiting till the holiday was over to get Jamie out of town. I spent this Christmas in a cheap hotel room getting drunk.

We said good-bye to all Jamie's derelict friends in Baton Rouge and flew off heading across the Pacific. By the time we arrived in Honolulu, we were already drunk *and* fighting. She was blaming me for taking her away from her kid and I hated her for all the times she had gone out with other guys. That evening the beatings began again. Aloha, Jamie! We would spend the next fourteen months of our lives drinking and drugging, until death do us part. There were fights that could have killed either one of us. We were getting kicked out of apartments, but couldn't stop fighting long enough to find a new place to live. One time she jumped out of the car and I was so angry that I ran over her. We finally ended up staying with Ron, a crewman from my fishing boat. The entire stay was just one big drunk . . . lasting for months.

I managed to get into a drunken fight with Ron. We had both been drinking, but I was helplessly drunk. I remember being hit while lying on the ground and no longer struggling. The blow broke my jaw. Later at the doctor's office, I took several stitches in a cut lip and my jaw was wired shut. On the way home from the hospital and for days to come, I vowed to get even with Ron even if it took years. I could picture an ambush of some sort, a surprise attack. I remember waking still drunk from the night before and driving down the road screaming and cursing at the top of my lungs . . . "I'll get you and kill you" . . . I was completely insane. I was running scenes and events of revenge through my mind. People would look at me as if I were nuts. And so they should have, because I was. I called my old sponsor and he suggested checking into a detox at a Salvation Army facility. I did exactly that, but after spending the night there, Jamie came and busted me out. We started drinking on the way home and it wasn't long till I was full blown again; if in fact, I had slowed down at all. Shortly afterwards, I unwired my jaw in front of the bathroom mirror with a pair of needle nose pliers. The result of this impulsive behavior was two years of grinding teeth and biting my tongue. I'm sure that my jaw never healed properly, but over the years the human body adjusts. I still bite my tongue every once in awhile when I eat.

At this point I had accepted that I was going to be the harbor drunk. This position was currently occupied by an old man named Les, but surely he would show me the ropes. I looked at what I had done in my life in order to drink and do dope, and a feeling of total hopelessness swept over me. The worst part of the whole deal for me to face, was that I had lived through it all and now was going to be . . . this time . . . the harbor drunk. Why couldn't I have died in one of those horrible car crashes? Why couldn't I have played the scene as a federal witness against the mob and been gunned down on the courthouse steps? Or maybe on a drug smuggling plane that crashed killing all aboard . . . buried by my peers with respect and honor, Little Italy style? I would have gladly accepted any one of these ends, but please, God, don't let me live to be the village idiot. If, in fact, God wanted me to be in hell for all of my dastardly deeds, I was surely there now, and had been there for quite a few years, but just didn't know it. . . .

The end to this housing arrangement was at hand. After another near fight over somebody drinking my beer, I packed everything I had and was getting out of town. Jamie, for once in her life, was actually on my side. She was great at taking the other side of the argument. As I look back, she was always on any side of an issue, but my side. She helped me load the boat and we left port heading for the Ali Wai Marina in Waikiki Honolulu, about a five-hour trip. I didn't even untie the boat that day. In defiance, I cut the mooring lines with the biggest knife on board, saying to myself, "You people will never see this boat again." At some time in the stay there, one of the harbor rats had said, "You don't own this boat, Forrest. This boat stays here in this harbor and you just pay the bills for awhile. We will be working the boat long after you are gone." I would have sank her before I let another one of those runts aboard her again.

We pulled into Honolulu Harbor about beer thirty that evening. I'll never forget that Monday night. After checking in with the harbor master, Jamie and I cleaned up and went out to a sports bar to watch the football game. Most games were spoiled in Hawaii because of the time zones. Someone would always walk in the bar and blurt out the score. I think that's what happened that night. In any case, Jamie and I weren't much interested and were playing shuffleboard on a machine. I wasn't very good at it. Some guys came up to challenge us for the table. We played, but I really wasn't into it.

The fact of the matter was that I had just left one part of Oahu Island on bad terms and had to see a boat broker in the morning about listing my boat for sale. Everything I owned was packed on the boat and I had no place to stay except onboard, which in fact was really uncomfortable. I was on the verge of losing everything over a lousy game of shuffleboard and here were two guys acting real sharp and flirting with my old lady. To tell you the truth at this point, I really didn't care anyhow, anyway, anymore. I think I said just as much to Jamie and made my exit stage right. She came flying out of the barroom, screaming and yelling at me all the way back to the boat. I was done, I couldn't and didn't even want to fight back. When I arrived at the boat, she was following me and jumped me from behind. I pushed her aside and she grabbed one of my rods and reels; totally destroying it and throwing it over the side. I restrained her enough to get her off the boat. She said she was going to get a cop and five minutes later, returned with one. She was so out of control, she started to attack me again. The cop knew right then and there, she was the culprit. She grabbed another rod and reel, totally destroying it. The cop had seen enough. He had her down on the deck in handcuffs, while calling for backup. He asked me what I wanted done with her, and I told him I didn't care just so long as she stayed away from me and the boat. She was released later and ended up staying with some guy on his sailboat that night. When I awoke the next morning, I threw all the beer over the side and went to have a serious talk with God.

3 The Unconditional Surrender

The beach was but a short walk away. As the surfers went by me that morning, I searched for a large flat rock that I could sit on and look out onto the ocean. I didn't know whether to swim as far as I could until I drowned or to pray. Before it had been in my mind to pray and that had failed, or I had failed, it didn't matter . . . it hadn't worked. Well, I had come to pray and so I did exactly that. The moment of truth went something like this: "God, I'm an alcoholic and addict named Forrest, and booze and drugs have stolen everything in my life from me. It has stolen my family, my wife, my baby, my home, money, cars and land. It has stolen my friends, my career, my girlfriend, my education, my business as well as all of my dreams and ambitions. Lord, it has stolen my soul. I come to You now. I know where to go to get human help and vow to do so, but I need Your help. Almighty, help! In order to beat this thing, I *need divine help.* I promise to You, God, as well as to myself, I never want to touch liquor or partake in drugs, ever again. Please, God help me! Please save this broken man here on this beach this morning. Amen."

I walked away from the beach and back to the boat. No burning bushes or bolts of lightning happened . . . but I felt pretty good. My feet were light and a smile showed. I cleaned the boat and presented a copy of the keys to the boat broker. It was getting to be eleven o'clock and I had a noon meeting planned. I looked up and saw Jamie walking down the dock toward the boat. On my way back from prayer that morning, the thought of my obsession for her was present and I knew that I had to let that go, and I think for the first time I truly let her go. She said something like, "Well, what will it be?" I commented that I didn't know what she was going to do with her life, but I damn sure knew what I was going to do with mine. I made her an offer right then and there. I was going to get straight no matter what and she could come along for the ride if she wanted to, or she could stay in the gutter. It really didn't matter to me. I started off to my meeting which was about a twenty-minute walk away. "Where are you going?" I didn't even turn around. I just said over my shoulder, "To a meeting."

"Wait up," she said, "I'm coming too." Bless her heart, she *was* coming too. At that moment, I knew I had let go of her for real, because it really didn't matter if she was coming or not. The point was, I was going . . . no matter what!

That first meeting back, some folks recognized me and seemed glad to see me back. Nobody knew Jamie and that made her angry. She showed her anger when they passed the basket around. "Oh, sure. Now they want your money." I was really happy to get back and somehow I felt the weight of the world was lifting from my shoulders. I had always come back remorseful, embarrassed, resentful, trembling at the word 'sobriety.' But this time it felt good. I don't know how to explain it. I guess it felt kind of like going out for an ice cream.

There was no animosity from the side-liners like there had been in my hometown. I had been in and out so often that back home you could almost hear them say under their breath, "I wonder how long he'll make it this time." "Hey look, it's Forrest. He's going to make another stab at it." Once there was a so called old-timer that said on the day I received a ninety-day chip, "Give him a beer. He ain't done yet." I thought how very cruel of him. Even though he was right, he didn't have to announce it, and steal the moment from me. I think I realized later that even though some of them have 190 years of sobriety in a lifetime, *that,* somehow, is the best that they can do.

Jamie and I picked up a newspaper on the way back to the boat to look for some sort of apartment. Places were so expensive in Hawaii and I was broke. I immediately sold my fishing tackle off the boat and we borrowed a car to go house hunting. Oh, I failed to mention that Jamie had totaled my truck one night and naturally, I didn't have insurance. So we were off in a three-hundred-dollar clunker to find us a place to live. One of the places we looked at was to split expenses with a guy who was just getting a divorce. The place was pretty nice, but first Jamie and I had to have a little chat about what we were fixing to get into. I told her that under no circumstances were we getting kicked out of this place and that the first time she decided to get drunk, it was done. Over. Finished. Get it! She agreed, but not more than a week later I came in from a meeting and found her falling out of a chair; drunk as a hoot. I told her to get her things packed and be out of there in the morning. We didn't even fight about it. She went to a meeting the next morning and came home vowing she would behave. I gave her a break and we were off and running. Within a month she had a job, riding a bus an hour each way to work. I was in school getting a charter captain's license and running bootleg charters on the side.

It took awhile and wasn't easy, but we were making progress. I found a new sponsor. He was an old seaman, who had a boat, and I started hanging out with him every day. He had thirty-two years of sobriety and I knew he wasn't going to get drunk. I wanted what he had: peace, serenity, dignity, honor in his head and love in his heart; a lifetime of sobriety. From all of the people I had known, nobody had exactly that length of sobriety. I respected the mere fact of his longevity and would discover the rest later. His name was "Kokko," and he soon became the dad that I never had.

This man never shoved the program down my throat. He never hit me up with the Big Book. Never represented the step Gestapo. He just let me hang out with him. When I asked for advice, he was there. When I got angry, he allowed it. When I needed a friend, he was a good one. But most of all, most important, he just let me hang out with him. Some days I hated him. When I say I hated him, I don't think I ever hated him. I just probably hated myself that day. Some days he would talk my ears off about nothing in particular or anything important, but I would make that day without drinking another day in the bank. The first year was a little rough and unstable. There were times of real insanity. There were times of real frustration. But Kokko and I made a lot of meetings and Monday nights became my best night of the week. I had made six months and my old sponsor, Pat, had nominated me to become secretary of what was now the world famous Monday Night Men's Stag Group of Waikiki. We got a lot of mileage out of the group's name along with some fun. The big meeting of the week was a speaker meeting on Saturday night that generally held some 200 to 300 people, and we would get in grudge matches over what group we were in.

The group did elect me secretary and said it was to make sure I was going to stay around for another six months. Everything with service in Hawaii was backed up with six months' commitment. I was honored and did what I thought was a bang-up job. That first year I put the "steps" on the back burner. I stayed within the framework of 1,2,3, and held everything else back until I knew I was in this for real . . . this time. I considered *for real* having one year of sobriety under my belt. I made that first year and it was the greatest thing that ever happened to me. On my birthday, I was chosen to tell my story in front of my group. I covered my "drunkalogue" and cried a lot. It was truly a great moment for me. A milestone . . . I'll never forget it and wish never to repeat it. I remembered that year when I came back to my group. I met an old-timer at the door and we sat down to talk before the meeting. I mentioned that somehow this time felt different. That maybe the miracle of the obsession to drink had been removed. That maybe the miracle of God was in me; that this time it was real. He looked at me and said "Ah! Shoot, Forrest, maybe this time you just have had enough." I told him I didn't know, but whatever it was, I would damn sure take it. You know I think that a big part of the key for my success in sobriety today is that I *just don't want to drink anymore.* Somehow, no matter what happens in my life today, I just don't want to drink anymore. That is pretty simple and really very easy. Can you imagine how hard it would be not to drink gasoline? Well, that's how easy it is for me not to drink alcohol. As simple as that.

Kokko would literally give you the shirt off his back if he had to. I wasn't sure if this was good or if he was just some kind of sucker who had been ripped off time and time again and hadn't learned yet. We had a few guys pass through at that time that took advantage of us. Once I sponsored a young guy who just got out of jail. I let him stay at my place and got him a job. Kokko let him use an old car and we got him into service at the group. In about three weeks time, he had stolen Kokko's car, ripped off the group's treasury and quit the job. I was so angry that I screamed at Kokko, "How could he rip us off like that, after trying to help him and all." My sponsor calmly opened the Big Book and showed me in black and white the specific instructions in the "how to's" and "wherefore's" of working with others. I was dumbfounded. "Why didn't you tell me this before ?" I asked.

"Forrest," he said, "everything you have ever learned has been through personal experience, the hard way. You never learned anything from a book. So I let you learn the only way you knew how." How incredibly wise and patient that was, I couldn't believe it.

Things had gone really well for Jamie, although, she was still riding a bus back and forth to work. By this time all the bills were caught up and we had a place of our own. It wasn't long before we actually bought us a new car, a brand new Ford Taurus. As for myself, I had my captain's license and a boat that I was running for a pretty good ole boss. She was a 46-foot sampan with a bit of history that is still current. This boat had captured the biggest blue marlin ever boated, a 1,830-pound Pacific blue that hangs in the International Trade Center of Honolulu. Things were going well as I built my reputation as a captain. Some days did go haywire. Such as when the old ship's transmission would stick in gear and I'd come in and crash the docks. Once, with passengers aboard, I ran out of fuel. There I sat, adrift about three to five miles outside Honolulu. By the time I got a boat to tow us back, the whole day was over and the passengers weren't amused, to say the least. As we made the dock, I looked up and there was the old man, Kokko, waiting to grab the line, lend a hand as always, and pick up the pieces. I tell you right now, with tears in my eyes, he was there a lot to pick up the pieces.

I got to a meeting that night and you could see pain all over my face along with a little scrambled egg to go with it. The topic that night was the third step, turning our will and life over to God. I didn't have much to say. Suddenly this member who was talking, turned to me and said life had been great ever since he had turned his will over to the care of God, but sometimes that wasn't enough. If you really wanted to have things work in your life, you sometimes had to make sure the boat was filled with *gas.* The whole place started to crack up and I started to laugh. I knew these men really understood me and knew exactly who I was. *Kokko had made it to the meeting that night about thirty minutes ahead of me.*

I have met such dynamic people in those meetings. It is just incredible how talented people turn out after the miracles happen in their life. I have met so many actors, actresses, writers, lawyers, doctors, clergy, bankers, judges, business people and laymen. All of whom have much to offer society in the common cause of helping the suffering soul. Once, in the beginning, Kokko called me, saying we had to make a 12-step call. I immediately hesitated with fear. "I can't help anybody! I only have forty-something days." "You have forty-something days and I have thirty-plus years. Who do you think this drunk can relate to the best? Go ahead, get in the car." I can't remember the details but, I do remember how high it made me feel . . . being able to help or offer help to someone else, lifted me three feet off the ground.

I went to the Hawaii State Convention that year. It was an incredible experience. If you've never been to a big convention, do yourself a favor and go . . . over 3,000 drunks were in one place. It felt powerful and carried an aura with it, making you proud you were part of being there. The main speaker that evening was so witty and amusing; he could have been a host on a late night talk show. In his "drunkalogue," he had been a practicing lawyer, with so many failings that his firm not only fired him, but were about to strip him of his *Bar* privileges, pardon the pun. He told some wild stories and some sad ones about the damage in his life caused by the disease. You know the kind: loss of family, wife, kids, career, sleeping on the street, the run-of-the-mill pain we have all had to cope with. Right! In his recovery part of the story, he shared how life got happy for him and how he found God. Today things were worth living for. When he got to the punchline, he broke his anonymity. In sobriety he recovered a career in law and later was elected as a local judge. He went on to become a judge on the Supreme Court of the State of California. He topped it all by saying, "With the power vested in me, not by the State of California, but by God Himself, I order you to go out and enjoy your life. And sentence you to live it to its fullest potential, fulfilling yourself with unselfish love. Be happy, joyous, and free through the Power of God, and forever dwell in His Grace, as you trudge the road toward the Kingdom of Heaven." I was floored, everybody stood up with a roar; a standing ovation. What a dynamic character. I knew that this was the life I had been searching for . . . for so many years.

Surrender! Surrender isn't hard, not if you want to learn and improve the life you have. Stop drinking! Stop drinking isn't hard at all. It's very easy to remember all that we have talked about and if I wanted to go back to that I could. Or look at the judge who has dignity, decency, respect, joy and love in his life today. You make the choice. Let's look at our alternatives. The alternative for me is death. I know that today and hope never to forget it.

When I was in treatment, they kept telling me to *surrender*. I really didn't know what it meant. Then when I got up to that halfway house in Minnesota, they even put a sign over my bed with just the word, *Surrender*. All my life I have been struggling and battling to be successful, to be important, to be useful, to be somebody. It was a constant, daily battle to win. I argued with waiters, cab drivers, teachers, cops. I was miserable, and whatever victories I had were shallow ones. I have heard AA speakers say that I drank in order to be someone, and became a nobody. I came into AA a nobody and became someone. I like that. That shoe fits for me.

Sometimes I still try to take back control of my life. Each and every time I do, it gets stinking real bad. So I am practicing the *art of surrender,* and I start each morning from my knees with the third step prayer. Later in my seventh year, I met a man who saw I was having trouble in this area. And at that time, my life was going over some very rough spots. He made me make a commitment to start my day like this every morning. I would do it for a month and my life would go well. Then I would abandon it and things would turn. I have given newcomers the method and we would make thirty-day commitments. After the thirty days, they would give it up and things would go flat again. I saw it in everything. Just like the Book says, "Half measures avail us nothing." So now I just do it. I have found that the prayer is not just words anymore; it has become action. In a sense, like exercising. If you did fifty push-ups every morning, it's not a big deal after awhile, but without looking, your body becomes stronger, firmer, and you feel better about yourself. It just becomes part of your life, a part of you. And suddenly after repeating my morning ritual for a time, I became fit. This is how it has worked for me.

Today I don't fight with people, as much, am not as violent, or provoked as easily, and when I am, it totally throws me out of kilter. I just don't want to do that to myself anymore. Today, I find that the more I don't want to do negative things to myself . . . the more I don't.

Before we move on to bigger and better parts of the story, I'd like to mention a few things that happened in these first two years of sobriety. First of all, there were more deaths of those that were close to me. One of the guys that I worked with on my first job in Hawaii was found dead on the back deck of the boat. He was sitting on the big heaps of lines, stiff as a board from rigor mortis with a needle still in his arm. I knew this stuff was going around the docks and I did try to help when and where I could. But it is obvious when somebody is ready, and the fellows around there definitely weren't ready.

Then my mother's dear friend and sponsor's son died. I know I have said that my mom had not drawn a sober breath, and in my opinion she hadn't. She has been in and around AA since the fifties, but these ladies in my hometown have allowed the prescription addictions to continue while each and every one of them received chips each year. The truth is that they protected and defended these doctors at all costs, just as an addict protects his stash. In my mom's case, she has been under doctor's care, in one form or the other, all of her life. She has been considered to be too far gone to approach. Her sponsor, who is a very good woman, along with my mom, and another woman named Jenoria, prided themselves as being the Three Musketeers of Woman's AA in Lafayette. Each of these women had alcoholic, addict children. Jenoria's daughter was hell on wheels. Katherine's son, Johnny, was as bad. And then there was me. The threesome were always wondering, "Who's going to make it? Which one would live?" Well, Jenoria's daughter, Judy, died while I was in treatment in Kansas. Katherine's son, Johnny, died my first year of sobriety. He had been around several times and had even picked up a chip for a year. But he went back out and his body couldn't take it anymore. He literally drank himself into the ground. So far, I am the only one to survive.

Next, there was the apartment manager where I lived. A nice, handsome, young lad, whom I knew only casually till I saw him one night at a meeting. He asked if he could come by and talk from time to time and naturally I said, "Sure!" That was the last I saw of him. He was found that weekend hanging by the neck in his apartment.

The point of all of this is . . . these people didn't surrender and it killed them. I work with a lot of alcoholics and have heard all the excuses. When I start hearing the excuses, I stop offering the answers. "Well, you know where to find us. Come when you're ready," I say. Many of them never get to the door. Some of them ask me to be their sponsor. I say O.K., but I want ninety meetings in ninety days, no relationships for the first year, and you will start working steps immediately. Here's my number, go home and think about it and give me a call. They never call back because they aren't ready. Nor will they go to any length.

I'll tell one more story and then we'll go on. This is more recent and just happened yesterday.

About three years ago, still living in Hawaii, I received a call from a guy I had known since childhood. He was calling me from halfway around the world. "I heard you got sober. I have heard you made it through and are doing good. Tell me, Forrest, how did you do it? I know your history and know about all the treatments. Which one worked for you? Was it the one in Kansas?" I replied, "No! All they told me to do in Kansas, after they took the money, was to go to AA and there I would find health." I spent at least an hour on the phone with him and then I started to hear excuses. Something like a patient paying the good doctor for advice, but already has all the answers figured out. Well, I told him to get in touch with a few good friends back home and gave him their phone numbers. "Make sure that you go to some meetings with these guys and try to get a foothold on this thing." I then called the guys up and told them to keep an eye out for him, lending a hand if possible. I hadn't heard anything more, till yesterday, Easter Sunday. I called my mom up to say hello and she told me the whole town was buzzing. Anxiously awaiting the news, like a gossiping goose, I was to be left in shock and anger. Seems this guy, I hesitate to mention his name, was sent to a doctor's office to undergo a medical evaluation to determine if he was to be sent away to the state hospital for the insane. At the time of the verdict, this crazed fool jumped up, and stabbed and killed the doctor. He was taken away all right and it wasn't to the funny farm. He's facing life in prison for second degree murder. Coming from a rich home, a professional father, famous uncles, and a good, reputable name, he had every benefit. But he would not surrender as long as there was an excuse left. I talked to my old friend Billy and he said that the guy had made a few meetings with him. But it apparently was not enough. The doctor left a wife and two beautiful daughters behind. I felt so angry, so hurt . . . so much loss.

I took the story to a meeting that night and spun it into a tale of gratitude . . . "There but for the Grace of God." How many times had I almost killed Jamie, or how about the guy at the condo I almost threw off the twenty-plus story balcony, or all of the times I was packing pistols and dealing dope, the fits of anger turning into a mad search for revenge, only to get drunk before the damage was done. I told the group that I was grateful just to find my way to the doors of AA, and if nothing more happened than my life being spared, I was eternally grateful. . . . More than enough had been given to me.

What kind of Easter do you think my friend had, sitting in his cell? Someone at the meeting responded to the story with the confirmation of his greatest fears . . . *along with mine.* To go back to drinking. To be bound for jail or an institution. To live and not to die.

4 Going To The Big Leagues

I had already started to get bored with all God had bestowed on me; sobriety, a good sponsor, a pretty neat job, and an apartment that I had actually lived in for over a year. Not to mention Jamie and I together, both clean and sober, working and being productive. I was ready to move on to bigger things. I stayed up nights thinking of ways to get back into business for myself. Since my first business venture with the boat had eventually failed, it was going to take some hard convincing to prove that I could be successful and stay sober at the same time. I knew I was ready. Staying up till the wee hours of the morning figuring out what boat to buy; what business would be best; the charter boat business or commercial fishing? Should I buy a boat in Hawaii and pay the premium price, or go to Miami and buy a boat right? There were hours of dreams about making the voyage from Miami and Hawaii in a forty to fifty-foot boat. Everything had to be taken into consideration: fuel stops, weather, pirates, food, crew. All these logistics had to be financially weighed; not to mention the risk of life.

My father came through for me and agreed to help me get a bank loan. My other boat had been sold and that note paid without too much damage to anyone involved. I asked him to please come over and we would pick out a boat together. And to my surprise, he came.

I showed him the operation that I was working for and the boat I captained. The boat was actually making a little money and I had the records to back it up. I could tell he wasn't impressed and was just going through the motions. After we checked him into the hotel, we had a moment alone in the room. I had checked my motives, and felt that it didn't matter about a boat. I had never had much time alone with my dad, so this was it. I started to pour out my heart. "Dad, remember when you told me to get an education?" He said, "Yes." 'Remember when you told me to stay away from any kind of drugs?" Again, "Yes." "How about, to keep moderation as a motto?" This time, "Yes. What's all this about?" he said. "Well, I just wanted you to know that you were right. You've been right about everything you have ever told me and I have been wrong. I've been dead wrong about everything all of my life. The whole world hadn't been out to get me and I have fought everyone and everything all of my life for nothing. I don't know where I took the wrong turn, but I'm sorry for all the damage done and the pain I've caused. I don't know how, but I'll make it up to you." He said it takes a big man to admit that. I stopped him right there, "I'm not a big man, I'm just wrong." He smiled. We hugged, and I left so he could rest from the journey. Whenever my dad and I hug, it's awkward and uncomfortable for me. So far that hasn't changed, but I'm working on it.

That night I made another deal with God. If He would be so kind as to see that I get a boat, I would stay straight throughout the venture no matter what. If I was successful again this time, I wasn't going to throw it away. I decided to go to Kona where the biggest fish and the best captains were available to compete against. If you were going to pitch in the major leagues, why not for the world series champions? . . . a dream come true for me. Once upon a time I used to sit on a bar stool and say, "When I make it big, I'm going to move to Hawaii, buy a yacht and fish for marlin." The problem was I could never get up from the bar stool, but through sobriety this dream was coming to be reality.

The next morning, my father and I, along with a boat surveyor, left for Kona, Hawaii, to buy a boat. We picked out a 44-footer, and had worked out a deal for not only the boat, but a charter-booking desk right in the heart of the tourist traffic. The deal was negotiated. I was to move over to the Big Island and be in business within thirty days. My dad was going to the bank to make another loan possible for me. After a good visit the rest of the weekend, my dad was off, headed back to the mainland.

I came home that Sunday night and sat down with Jamie, to tell her about the chance I had to go into business once again, this time on the Big Island. She looked at me and out came these exact words. "If you think I'm going to follow you all around the world buying boats, you're nuts." I had remembered seeing a bumper sticker that said "My wife said if I go fishing, one more time, she was going to leave me. I sure am going to miss her." Those were my sentiments exactly. A month later, I was living in a small rental house on the Big Island and was single.

The forty-four-foot boat and the charter-booking desk had fallen through. Here I had moved to a new island and busted up whatever remnant of a relationship that I had for nothing. The seller of this enterprise turned out to be a snake and even tried to keep my earnest money. With a hard-fought battle, I received all my funds back and after about a month, I had started on a new deal.

With a waiting list for space in the Hawaii state-operated harbors, there was all kinds of tactical maneuvering going on. Lawyers loved it and state officials smiled on the way home from work every time a transaction took place. Figure out the undercurrents of that statement for yourself. Valuable time was being wasted at my expense. It was weighing heavy on my serenity and my pocketbook. I remember making a lot of meetings and I still enjoyed getting acquainted with all of the new people in the fellowship. Sometimes when my days were going too bad, I was making two or three meetings a day.

To give you a picture of the island, or for that matter, Kona, you should have a little knowledge of Hawaii. All of the islands are uniquely beautiful, each in their own way. The Big Island was exactly that, the biggest island of the chain. It had two huge mountainous volcanoes that were still active, and Kona was located on the leeward side of the island, with a population of roughly 20,000. Kona did not have a lot of entertainment: a movie house, a bowling alley, and a newly done Municipal Park offering a baseball diamond and soccer fields, was about it. In wintertime the mountains would be snowcapped; truly a sight to behold while lounging in swimsuits on the beach below.

The town being so small, took about three meetings and everyone knew you. We had a place called the Pavilion, which was outdoors in the churchyard. I used to complain about holding meetings in such a barnyard atmosphere, but most of the visitors from the mainland loved it and so did the old-timers who had founded it. Every once in awhile, a rat would run across the rafters during our eight

o'clock evening meetings. I used to laugh and say that I wondered how great visitors would think of this place if one of these rats fell into their laps.

After awhile I learned to love it myself. I spilled my guts the first few meetings, all about going into business, and people would share their program with me. Six months after my dad made the trip to help me, I learned finally how to let go of things. Wouldn't you know just as soon as I did, the deal was completed and I had a boat and a charter cooperation. The torment I had to deal with just to go into business, hopefully, wasn't going to be the measuring stick for things to come. There are always people around who spoil things for you, and lo and behold, my boat slip was right next to the most arrogant, egotistical, and boastful boat captain you could ever want to meet. You could count on them to shower on your parade no matter what. It never fails. I was investing $150,000 to go into a business for myself, and the guy next door was making me miserable on a daily basis.

The boat itself was enough to take the sting out of things for awhile, like materials things can do until the newness wears off. She was a 36-foot Hatteras, sporting a fly bridge and tuna tower . . . my "dream boat." Although not as large as I once thought I wanted, she proved to be cost efficient and maintenance-free in the long haul. On the sea she proved herself a worthwhile friend and gave me a lot of confidence. It was great and I started catching fish right away. Soon the charters were coming my way and I was breaking into the fleet. This, of course, was a big relief, since I was under tremendous pressure not to fail; tremendous pressure, mainly from myself.

In less than three months, I had caught my first real big fish, a 662-pound blue marlin, and I was really stoked. The client was a Canadian and he didn't even tip me or the deckhand. In this business you soon learn little things like . . . the Canadian tourists are the worst tippers, so you give 'em the service cheap tippers are supposed to get. Not long after getting my feet wet in the biz, I learned how to cut corners and how to make efficiency work for me. I believe in treating people exactly how they treat me, and I got along fantastically with most of my customers.

My deckhand was a story all by himself. He just walked up to me one evening on the docks while I was admiring a 60-foot Hatteras that had just pulled in for the night. During our conversation, he asked for a job and proceeded to show me his last quarter from his pocket. I asked him if he knew how to fish, and he said, "Oh yes!" He had fished in some of the big tournaments on the island but as a client, not a crewman. Turns out this guy was a down-on-his-luck veterinarian from California. Seems that the ex-wife took all the money, the house, the car, and the kids. I sure could relate to that, and I told him to get his gear and come up to the house. There was no use for him to sleep on the ground. What a swell guy! He even knew how to play Hollywood-style gin rummy.

He and I broke into the business together and we lasted almost two years. On the right kind of trips with the right kind of people, I would let the charter customer drive the boat while Doc and I were down in the cabin playing gin. We caught some nice fish together and had some good laughs. Once a fellow came over from the Honolulu AA, needing a place to stay and sober up. We made a deal for him to move in with Doc and me and the three of us would share the expenses. We got along pretty well. He and I made meetings, but we never did get Doc to go to any with us. It turned out to be fun for everybody. We even had nickel, dime and quarter poker nights. Doc and I had a betting thing going. We would put a dollar on just about anything: baseball games, golf, tennis, water polo . . . it didn't matter. Lee, the new guy, didn't do much. Doc and I would go out on fishing trips and get back home to find him asleep or playing solitaire. One time, driving up the hill, I said to Doc, "A buck: Lee is sitting at home asleep in the big chair with the TV remote control in his hand." Doc answered, "You're on." I won the bet. I was telling another friend, Dayton, about Lee, and he said that Lee was waiting to get drunk. I had never heard that one before . . . he's waiting to get drunk. I didn't understand, and Dayton said, "You kick him out or you get into a quarrel or he slams his foot in the door of your house; it doesn't matter what, he'll use it as an excuse. He is waiting to get drunk and it's going to all be your fault. He got drunk and used a relationship for an excuse.

I've seen more people get drunk over relationships than anything else. Again, I once heard it said that second to the craziest thing you can do in this program, is to get in recovery and start a relationship right away. Once at a meeting, I made the profound statement, "Coming to AA and using the program for a social club is like going to the nuthouse to pick your spouse." You can imagine how much they loved me at that meeting, especially the married ones sitting there hand-in-hand. It is the truth and if you don't believe it, you had better go back to the second step; you may have missed something. The most frustrating thing for me to watch is the women coming in for the first time and getting hit on as they come through the door. I think it would be frustrating to be a woman in any case, having to tolerate this sort of thing every day. Please give them a chance to at least come up for air and allow them a breath of sobriety before you make an emotional nutcase of yourself.

Nevertheless, it is a two-way street, and I have seen some positive results with some of it. At least it keeps them around long enough to buy some time. My biggest hero in the program once said that he made the meetings in the jails just so he could smell the perfume. So I concede there is a place for it. The bottom line being that all of us are uncomfortable in our own skin and have been looking for an external *fix* all of our lives. We have tried to fill the hole by abuse of alcohol, drugs, sex, relationships, gambling, money, careers, material gains, and anything else that would give us a moment of escape. The key word being *escape*. You must realize that a new relationship is exactly that, escaping from one's self and occupying ourselves with the external fix of the companionship of another human being. When I mention an external fix, it goes into all sorts of tributaries of co-dependency. We need this relationship to boost our self-esteem, prop up our ego, fill the void of boredom, loneliness, restlessness and discontent. Some of these affairs get into the area of . . . "Do anything to me, just don't leave me." You can easily see how unhealthy that is. My relationship and marriage with Jamie had a lot of love/hate characteristics. I have seen many of my friends put up with just about anything in order to stay in a sick relationship, just so they wouldn't have to be *alone*. In fact, we all finally reach a point where we realize that it is a sick relationship and we want better than that for ourselves. Even if it means being alone. Regardless, some of us wait to get drunk because of whatever the reason.

Meanwhile things were going well down at the Pavilion and also in my business. There was a lady that would make about thirty days at a time, and go fishing, come back in and pick up another desire chip. This went on every two weeks for quite awhile. She finally would up in treatment. The group was sending her a card and when it came to me, I wrote the message: "You can go fishing without drinking. I do it all the time." And I signed it, Capt. Forrest. Now I don't take credit for her sobriety, but she hasn't had a drink since. She says that she thinks about that card all the time and I really did help her. She grew into such a nice lady. It's wonderful to see people grow in sobriety. Some folks learn quicker than others, and her recovery came quite fast. Her name is C.B. and she still loves to fish, and she's still sober.

We had a lot of talks at the Pavilion and I left a lot of people back there whom I really love. I felt I was home and in the Big Leagues, finally competing for business, tournaments, and life.

At one of the meetings, my anger was showing up again, so we made it the topic of discussion, as it had often been in the past. Many times we hear things and never think too much about them or can't use them in our lives yet. Well, the group was talking about anger and once again the subject of fear came up. I had often heard it stated that underneath all anger lies fear. I've always thought, "Yeah, but I'm not scared of anything." I've stared death down and made it run and hide. I had beaten this alcohol thing and been through countless periods of pain. Fear! . . . I fear no one! Then this little old lady said, "You know, I have fear and I have found out that my biggest fear is the fear of looking bad." All of a sudden the lightbulb went on in my head and I knew I had fear and plenty of it. I can't tell you how much this one person's reflection helped me. It opened the door to new heights of surrendering. Many years later, I learned that under all anger **is** fear. Under all fear is hurt. And under all hurt there is loss. It gets down to two very primary losses, a loss of self-esteem, and a loss of love. In my case, all of this anger that I have carried around with me all my life was from a loss of self-esteem and a loss of love. Yes, that's right. It made perfect sense, too, because I have suffered immensely from both problems since I was a little kid. I think that underneath all of our addictions, we all suffer from a lack of love.

Knowing the answers doesn't always solve the problem for me.

It does give me insight into how and why I have been the way I have . . . behaved as I have . . . become who I am. It gives me some direction in working toward recovery. I would not work to control anger, but to control my fear. Instead of resenting my parents for their failure to nurture me and their continuous destruction of my self-esteem, I should strive for forgiveness, and try to learn to love myself. There are so many things that I have heard and still hear at meetings, but am not ready to use. For years I was told to pray for the people I resented, but it took years to get there. Sometimes it is still very hard to do and I have to suffer the pain first. If you experience enough pain, it will motivate you to do anything; even to pray for someone you hate.

Many times I got angry at meetings when it was changed from an AA meeting to a marriage counselor's office. How many times have I seen people bring in their wives to show their "Honey" how well they are doing. "Look at me; I'm picking up a sixty-day chip." Things get so complicated and much of the rationalization and justification comes into play when involving love and marriage. One of the worst cases I have ever worked with was a guy who would just get loaded and wind up in intense trouble every time. The family asked if I would go to the jail to talk to him. They put us in a holding cell and he looked up at me and said in a teary little voice, "Everything would be O.K. if only I had my wife back." I picked up the chair that I was sitting in and smashed it against the wall. I'm lucky they didn't put me in jail with him. The point is, that it gets in the way of people understanding the real problem. In this case, this person got out of jail, remained in the program and started a new relationship immediately. After six months, they broke up and he was frantically searching to replace her. He is one year, three months old and trolls the streets and bars 'til two or three o'clock in the morning looking for action. His obsession with women is so intense that his recovery program, in my opinion, is at day one or less. If he doesn't have a spiritual awakening soon, he won't make the end of the month. He walked into my house the other night and said he needed to be married and everything would be O.K. . . . exactly my case and point. Sexual addiction now defined. Back to day one.

If we don't deal with ourselves, we will have to face the dragon sooner or later, but we will have to face it. They used to tell me, "Forrest, sit and be still. God can't hit a moving target." Telling me to be still is like telling me not to breath, or like telling somebody not to worry. Yeah, right! In fact, if we are always looking for that external fix, we are blocking the light of God's Grace from reaching us. Just think how much faster Johnnie can run and how much higher Johnnie can fly without the excess baggage of being addicted to drugs, alcohol, sex, gambling, material things, relationships or what have you.

I can't really say exactly what the second thing is that makes people go back out and use: whether it's anger, fear, loneliness, boredom, or what, but every time they come back to the program, they always say, "I quit going to meetings." They say it every time; it never fails. It's always a thousand different reasons why they quit going—personalities before principles. I got too busy, started a normal life, had to take care of the kids, or whatever. The point is that they got drunk, are back to day one, and lucky to make it back alive. I saw a lady with about nine months' sobriety, go back out and drink. God bless her. Darn, if she didn't come back about a week later crying, shaking and scared to death. She made about ten days and died over the weekend in a fire. She passed out drunk; her cigarette caught the bed on fire and she burned to death. We lose our priorities and drink again. Some of us, like me, just don't have another recovery left.

Where I now reside, I am often the senior member of the group. It had a very scary effect on me when, at one of these meetings, the topic was, "Does the program work, and if so, how does it work in your life?" After all the usual unoriginal boring rhetoric, I said, "I don't know if the program works. You tell me. How many of you, in this room tonight, have more than five years of sobriety." Oddly enough, but expectedly, no one could raise their hand. "If this thing works so doggone good, where are all the old-timers?" In principal, there should be all old-timers here with only a few newcomers. Why then, is it always the other way around? I often think about the time I went to my first state convention in Honolulu and they asked everybody to stand up. There were about 3,000 of us in the main ballroom and it made quite a commotion. They then asked for you to sit down when they called out your number of years of sobriety. After the first year, only half of the room was left standing. After five years only about 15% of the room were on their feet. When they arrived at twenty years and over, only two people were left on their feet and one of them was my sponsor. Somehow it just seems like that is backward.

I have a friend that turned five years' sober about a month ago and didn't pick up his chip. I asked him what he was waiting for and he said that he would wait and get his chip from his home group back in Virginia. "You are living in Alabama now!" I replied. He said, "Yeah, but I'll wait and have my sponsor give it to me." I told him, "Hey, look, the chip is not necessarily for you. It's for everyone else to see that there is success in and around the program." He started to argue so I dropped the subject and suddenly remembered what happened to me at my five-year anniversary, which I will share with you later.

People look to the old-timer for guidance, direction, stability, and proof that it works. Sometimes it's overwhelming. Once I chewed a meeting out for not starting on time, not staying on the topic, picking a chairperson five minutes before the opening prayers, so on and so forth. I said I'm from the old school where a good meeting is one that starts on time and ends on time. People are supposed to have chairpersons or secretaries, in committed service positions to show up and start the meetings. The other day we had a topic of self honesty and it got a lady started who spent twenty minutes talking about her eighteen-year-old-son wanting to marry the first woman that laid him. It then went to the next lady who talked another ten minutes about fifty-cent Rice Krispies coupons that she cheated with at the grocery market. I walked out. I spoke of the format being wrong to go "ring around the rosie," with everyone getting a turn to share. Many of them do not have anything to give; The Rice Krispies coupon being a prime example. Everyone need not share at the meeting. Suddenly, a newcomer started to cry and said, "We don't know how to run a meeting. Where are the ole-timers to show us?" Boy, did I bite off more than I could chew. I recommended to at least start the meetings on time and bring up the rest at the next business meeting. Furthermore, if I had any complaints, I should also attend, but since I am not willing to attend the business meetings because I have had some horrible experiences there, I should not have jumped up and complained . . . especially when I was not offering any leadership. Mr. Controller, here I go again!

The next week everyone looked at me like Mr. Bad Guy, the heavy, the sobriety police. The meeting got out of order, as would be expected, with a chairperson with less than sixty days of recovery and I walked out. I realize I'm not a good example of gratitude, but when things start to make me want to blow up, I have to leave. Maybe that's where the old-timers are. I've heard that they just get tired of putting up with it. I'm sorry, but if I don't make meetings, I'll get drunk. And if I start to resent the people, I just have to pray for them like they taught me.

As the months went by and I got my business venture on its feet, the newness began to wear off and the boat started needing repair. One morning, on my way down the hill to the harbor to fix the boat, I was so angry that I was cursing the boat, the business, God and everybody else. As I pulled into the harbor and made the turn to the boat slip, I saw a man walking down the road with a bucket full of clothes, apparently on the way to wash them at the nearest faucet. The man had a smile on his face and that was a hell of a lot more than I had that morning. When I see people less fortunate but happier than me, it makes me feel like such a whiner . . . a spoiled brat that had the whole world by the tail and did his best to screw things up. Today I can recognize this as just more of beating up on myself; a continuation of the mental beatings inflicted by my parents. I can accept it and move on.

Not too long after that, another one of my harbor neighbors, Doc "Kid" McCoy, a transit from Texas, and I were talking and getting to know each other, when the subject of gratitude came up. I had mentioned the incident of the man with the bucket and how I had been taking Hawaii for granted lately. When Doc and I started talking about his life, I learned that upon his arrival here in the fifties, he bought a whole mountainside at a good price. He has lived and loved it here in Hawaii ever since. Even though he travels extensively he is always glad to get back to the islands. I asked if some of his mountainside was for sale. Doc told me he had sold some acreage for $100,000 an acre, but had to quit because he was creating a huge tax problem. "Good God! How many acres have you sold, Doc?" He said about a 100 or so. Wow! Doc had $10,000,000 or more in property sales. That was it. I was impressed . . . not with his money, but the way it hadn't changed him. He was just as down to earth country as you ever want to see. Doc's greatest pleasure in life was working with kids. He had a retarded son and felt fortunate to be able to help retarded kids in any way he could. Doc told me about two Filipino kids that helped him out once when he needed help, and he was putting them through college now. You could see his pride when he said they were still in school today and how he enjoyed watching them grow. I walked away from the harbor that day dumbfounded. Here was a guy that wasn't in the program and he knew the secrets of life and how to live them. He knew what was important in life and how to manage his ego; not to allow it to interfere. I was impressed indeed. You should have seen his boat. It wasn't much to look at and I'm sure Doc could have afforded any boat he wanted. He and his friends used to laugh and have a lot of fun with that boat. It wasn't about ego. One morning I arrived to find that his boat had almost sunk. I jumped aboard and got the pumps running. I must have pumped her for hours and when Doc got there, I told him we had almost lost her. He laughed as if, "not to worry" and said, "Oh shoot, Forrest. That ain't nothing. She was on the bottom when I bought her."

Not too long after this we had a bad storm in Kona and all the boat captains were down at the harbor tying up their boats and making sure they would be secure overnight. When Doc McCoy got down to his boat, I had already done his tying up, but we double-checked it. The wind was blowing hard and the rain poured down. We hurried to finish, said our goodbyes and left. McCoy got up to his house, slipped on the wet tile on his lanai, hit his head and went into a coma. The good man never returned to consciousness and was dead two days later. I couldn't believe it . . . life is so fragile, I thought. Here was a man with all the money in the world and really had things figured out. He could have taken it easy, but was out helping people most of his life. And after all of that, ended up dead from a fall on his patio. I don't know why, but it left me feeling powerless over everything.

We took his ashes in a caravan of boats and dumped them into the ocean. It is a beautiful way of doing a burial. The boats follow each other in line to a spot and then they back into a circle while the Reverend says a few prayers. The ashes are scattered into the ocean followed by flowers. Most mourning happens here as each person places his flower leis on the water. It always touches you when you hear the crying and sometimes even screams of sadness. The boats proceed in a circle, making three passes around the flowers and burial site; then head for the harbor. I always leave wondering, "Is that all there is?" Still I have often thought this would be the way I'd like it done for me when it's time. Soon things were back to normal and running smoothly . . . and for me that meant uncomfortable boredom. That old feeling of loneliness came back and I started reaching for the phone to call Jamie in Honolulu. I fought it back, but it kept coming at me time and time again. I had to make at least one more call . . . to finalize things and be satisfied in my mind . . . *one last time.* She answered the phone and, to my surprise, was glad to hear from me. We made a date and started to make plans. Soon we were married. It was a real nice ceremony at a little church with a beautiful Hawaiian background. It was really what we both wanted and I know that we have and will always love each other no matter what.

The roommates were all gone not long after this, and Jamie and I were together living in the small rented house in Kona. We were both working and doing well. Our lives were really coming together and prayers were being answered. Jamie's daughter, lost in a custody suit, was coming over to stay with us; this was major progress. Before long we had enough money to look for a house to buy.

My dad made a trip to see me and brought my sister and her kids. My sister was going through a divorce, which was to consume her for years to come, and my dad was trying to ease her pain. The trip would be a nice release, but it didn't work out that well for her. She spent her entire time in Hawaii drowning in pain. My dad stayed over after she left and helped us get a house. At times it's hard to believe how much he has helped me when I think about the other times when he has put me out in the cold like a dog. Underneath it all, I think he loves me very much. But things just don't always work out *exactly* according to the rules, or how I wanted them to work out.

The most incredible part of it all was that here we were living in Hawaii, making a good living, and yet there were still times of being restless, irritable, and discontent. It wasn't even six months when Jamie and I were fighting again; screaming in each other's face. One day I just stopped in the middle of it and looked around us. I said, "Here we are in Hawaii, both making good money, living in a $200,000 home, a $150,000 boat in the harbor, two brand new cars in the garage, and we're yelling and screaming about a bill that needed to be paid and would get paid, or a dish left in the sink to be cleaned that would get cleaned. We've never done better and it's never good enough." Boy! I guess you could write a book on that if you wanted to. I guess it must be human nature. When happiness seems to put us on a new plateau, we quickly become dissatisfied with the status quo and want more. We become resentful of all the little insignificant nagging things that haunt us and have nothing to do with our discontent.

Despite all of the distractions, my business was prospering, and after a few years I was getting bookings in the big tournaments and having pretty good results. I had taken second place in the Lahaia Jackpot on Maui and came in first place in the Hawaii Big Game Juniors Tournament. We had been on the circuit for some time now and starting to get lucky. Soon I would be catching top fish in the world famous Hawaii International Billfish Tournament; making the papers and TV. To some clients, maybe we hadn't completely

won the tournaments, but they wouldn't have traded their captain for anything in for the world . . . and to me that's winning!

I have yet to know a fisherman that couldn't keep your attention for a few days telling fish stories. I won't keep you for days, but here's a few quick ones. I had three very enthusiastic fishermen one morning, and after briefing them and going through the drills of what to do when we had a strike, I answered what seemed to be an hour's worth of questions. They finally calmed down and I settled into hunting for fish. Suddenly this large brown spot came to the surface and I turned toward it to get a closer look. As the boat got close enough to get a look, I saw that it was a giant manta ray, weighing 3,000 to 5,000 pounds. I immediately started to turn the boat away from it, but the lures weren't going to clear the fish and we hooked the ray in its giant wing. Snap, popped the rubber band which held the line up in the outriggers, and the huge Penn 130 H Reel started to scream. Now, because I already knew that this wasn't a game fish, I didn't get excited and my first thought was to cut the line so I didn't lose the entire spool of line.

As I turned and looked at the back deck, the charter was very excited and the chosen angler was already in the fighting chair waiting for me to gimble the rod to him. I figured what the hell! He doesn't know what kind of fish it is so why not give him a rush! The rod was bent down pretty good and the drag screaming loud. I looked at him and said, "Are you ready?" He said, "Yeah, give me the pole." I pulled the rod from the rod holder and set it in the gimble of the fighting chair. Even though he had a good grip on it, when I let go, he came out of the chair. The force of the pressure on the rod stood him straight up out of the chair. I grabbed the rod and went back to the rod holder. The drag on the reel seemed to almost be smoking and I needed to do something fast or I was going to lose about $100 worth of line. I looked at the angler and this time shouted, "Are you ready?" He braced himself in the fighting chair, locking his legs up against the transom of the boat. "Yeah, give it to me." I gave him the rod and he came straight up out of the chair and would have gone over the transom had I not grabbed him for dear life. At about that time the line snapped going off like a .22 shot. I commented, "Well, that's it. We broke the line." He had eyes as big as silver dollars and his knees were shaking. He asked, "What was it?" I answered, "I don't know, but it was a big fish, that's for sure." I turned and set the lines back up and we didn't catch anything that day, but he'll never forget that trip. I guarantee it.

I used to get a lot of Japanese charters. They never could speak English so it was especially hard to land big game fish when the anglers couldn't understand instructions. Most of them made crucial mistakes that either cut line or released the line free; either way losing the fish. Once Kokko and I had four Japanese on board. I was down on the deck setting out lines and Kokko was on the bridge driving. After setting out three lines, I was setting the fourth and a giant marlin came in and ate the short corner lure. The drag was smoking as the line peeled off the reel and the fish started to put on a magnificent show. All 600 pounds of him was jumping full body out of the water and we were being spooled fast. I screamed at Kokko, "Slow the boat!" Nothing happened. I yelled, "Kokko, stop the boat!" Still nothing. The great fish finally threw the hook and upon getting away he took off toward the starboard side of the boat, greyhounding in and up out of the water until he passed us up. It was truly magnificent and I was yelling at the top of my lungs, "Kokko, look at the fish. Kokko, do you see the fish?" Still nothing, so I looked up at the pilot's chair and the boat was being steered by automatic pilot and no Kokko in sight. I panicked. I thought the old man had a heart attack. Surely something had to have happened. No one could have been aboard and missed all of the commotion. I ran up the ladder to the bridge and Kokko was all the way in the front of the boat with a pair of binoculars glued to his nose. I yelled, "Kokko, what are you doing?" The old man turned around and said firmly and with a tight cheek, "I'm looking for fish." I almost fell down laughing. He missed the entire show, which was well worth the price of the ticket. In an illustrious career, I was to deck several marlin over 600 pounds, a few over 700 pounds, missing two over 1000 pounds and catching sharks well into five figures.

My biggest yellow fin tuna was 225 pounds and I have had all five trolling lines go off at once with both marlin and tuna. The fishing that I liked best were the trips with the most laughs, having nothing to do with good fishing, but good company, good people, and good fun.

Remember my friend, Dayton, who I mentioned earlier in the chapter? Let me tell you how we met. I was taking a fishing charter out when I first got started in the business and had just gotten my captain's license. These guys were all in the program and I was using my sponsor, Kokko, for a deckhand. After trolling lures for at least half a day and no bites, I happened to look back into the pattern of lures and we had one hellish tangle. I let Kokko take the wheel and started to clear lines to get to the "bird's nest." It was not a pretty sight and by the time I piled all of this mess onto the deck, I was turning a pale shade with embarrassment. The body English was enough to keep anyone at a distance for quite some time. At least long enough so I did not have to make explanations to anyone. All of a sudden this Hawaiian guy popped up and said, "Oh! Look how angry this guy is. I've got to meet this guy." He came over to my side and asked if he could help me untangle the lines. You can imagine what I wanted to say to him. Only I didn't dare 'cause he was bigger than me. By the time we had untangled all of the mess, we were laughing and joking and both knew that each of us had crawled out of the same identical "hole."

I've now known this man for eight years or so and he came all the way from Hawaii to the mainland to visit me where I live now. Please, you have to understand . . . while I was drinking, I was the guy getting kicked out of my family, out of cities, states and countries. To have anyone visit me at all is a gift. I still have family and friends who haven't yet made the four-hour trip from Louisiana to my house and this man came all the way from Hawaii to visit me. That truly says a lot about how I've changed. We took a trip in my new car, covering a four-state area and had a wonderful time. At one point during our trip, Dayton said, "You know, if I would have written down that morning I met you what I wanted from that fishing trip, I would have written down something like a 500-pound marlin or a good catch of some fish. Instead I got a friend for a lifetime, and someone that knows me right down to my very soul." The events and the secrets we have shared are immeasurable. I asked him at that moment to promise me that when I died he would see to it that my ashes would be spread outside the boat harbor at Kailua-Kona, Hawaii, a very special place for me. He said he could do that . . . and we dropped the subject. This is the best example I can give you on the topic of what sobriety has done for me. I've heard it said a thousand times. . . . "If I had written down what I wanted out of sobriety I would have sold myself short."

I can think of all sorts of ways that sobriety has far surpassed my expectations in business, such as the boat; in relationships, such as the one with Dayton and Kokko; to my homes, my place in Hawaii and where I live now and on and on. Now that I am conscious enough to go with the flow and not destroy things, it is incredible to count all the material benefits that have come my way. Trips, vacations, conferences, and opportunities keep happening for me if I just stop running from what God is trying to give me. Even with the sum of all these events in my life, there are still moments of weakness. I still wake up and doubt whether or not I am on the right track. I sometimes look at friends, their lives and my own life to see only negative things; forgetting that I am on borrowed time to begin with. I start to scheme devious things in my mind to recapture the thrills of the old days. The euphoric recall of being somebody way back when my hero was a gangster. I start to plan capers and scams to make a lot of dough to put me back on top. There I am back in control again. Once I was traveling first class on an airplane going to Las Vegas for a weekend. I looked at the guy sitting next to me and he was dressed in a suit; very clean and businesslike. My thoughts were . . . Look how successful this guy is . . . not realizing that I was the one with a briefcase full of money going to Vegas. The poor stiff next to me was on his way to L.A., ready to battle with a traffic jam. The fantasies are short-lived and the realization of, it just doesn't work for me anymore, returns. The phone usually rings about that time and it's a friend from the old neighborhood, who's also in recovery, and we just have a good laugh at ourselves.

While my friend Dayton was here on his visit, I took him with me to a mental health facility where I am a volunteer. The topic for the evening was man's manipulation, control and violence over women. Great stuff! It was a real powerful group that night with lots of participation. Toward the end of the group meeting, I mentioned that I had tried to fill the hole that I have in my soul with alcohol, drugs, money, gambling, women, material goods (such as houses, cars, and toys of all sorts); none of which have worked for me. "Through this program I have turned to God; learned to love God; found that God loves me, and through it all learned to love myself. By loving myself I have found that now I can love you."

Business dispute may turn boycott

By DAN BREEDEN
West Hawaii Today

A ground swell of negative feelings toward business changes at the Kona Marlin Center has resulted in the Hawaii Fishing and Boating Association calling for a boycott of the center's weigh station and store, and petitioning the state to allow a second fuel dock to be constructed in Honokohau Harbor.

With more than 100 members, the seven-year-old association represents the majority of Kona's fishing charter boat operators, according to president Pat Nelson. More than 60 people attended an association meeting Monday night to discuss problems members said they are having with the Kona Marlin Center.

Kona Marlin Center owner James Dahlberg said he did not attend the meeting because he was not given an agenda so that he could be prepared for questions.

"Despite several requests, I was not given any agenda and so I did not attend the meeting," Dahlberg said.

"The persons who called the meeting did not have the courtesy to contact me by letter or otherwise to discuss the rumors that have been circulating," he said.

Although most boat operators said they will continue to refuel at the Marlin Center, there may be a mass return to the Kailua Pier to pick up and return passengers and to have fish weighed on the scales there. Many charter operators have vowed to no longer solicit the Marlin Center's weigh station, sandwich shop or convenience store.

Phil Parker, who operates the Kona Activities Center, said at the meeting he will continue bringing his passengers to the fuel dock through today simply because plans have already been made. After that he will deliver his people to the Kailua Pier if that is what the boat captains want.

"I've got to go where the people are," Parker said, referring to the charter boat captains.

Parker's operation booked 72 fishermen on boats leaving the harbor yesterday, according to Parker.

The option to redirect captured gamefish to the pier is open to any boat operator, according to Harbors Division District Manager Ian Birnie.

"In fact, that was the only place available until a few years ago," Birnie said.

But, fishermen will not be allowed to book customers from the pier, according to Harbor Master Mike Silva. They can book them from boats tied to the pier.

Most boats will still be

See FUEL DOCK
Page 4A

—WHT photo by BOB FEWELL

FUEL UP, NO FISH — Boats line the dock at the Kona Marlin Center. A fuel line is extended to one vessel, but there aren't any fish hanging. A business dispute has charter boat captains thinking of boycotting the Honokohau Harbor business establishment.

Later, Dayton laughed and told me that I was not the same guy he met on the back deck of that fishing boat in Hawaii, untangling lures. I had grown a lot and that growth feels good today.

I could go on forever with helpful hints I have discovered in the program. Some can get you to sobriety quicker and some might get you drunk. One guy who was doing a lot of slipping, asked me for advice and I told him to go out drinking and do a good job. I told him my last drunk was fourteen months long. It did the trick and I had enough that time. He took my advice and came back in to say, "But, Forrest, it almost killed me." I said, "Good, you are lucky. You made it back." He has been sober ever since, but it really **could** have killed him.

Surely, I had all that I had ever dreamed of and, at least by my standards, had arrived in the big leagues.

Later, as we pursued our lives, both Jamie and I drifted away from the program. She had become branch manager for an electric firm and I was interested in making more money in my business. Sometimes she would leave the house at six o'clock in the morning, and not get home until seven or eight at night. I was doing all I could to compete for business and many times it led to a political power struggle in the harbor.

We even formed a Captains Association and boycotted the harbor fueling facility whose new owner was really sticking the entire fleet on all sorts of issues, mostly price gouging. I was taking a stand and being quoted in newspapers . . . *leading the cause of justice.* There was talk of multi-million dollar antitrust suits against the state for forming a monopoly in the state-owned harbor and awarding concessions to the highest bidders for fuel. It became very apparent who controlled things at times like this; who was playing child's gossip games; in fact, who had the backbone; and who was a whore (so to speak). I was sticking my neck out and everybody was letting me do so, on their behalf. I didn't get my head cut off, but things in the harbor had changed. People were different and I had taken everybody's inventory. Somehow it never felt the same. I had won the battle *and* the war, but things had a different taste now. I had lost respect for captains that I once looked up to, lost respect for how the entire harbor operated, and the charter captain profession as a whole.

I walked into a meeting one night and there was this old biker friend of mine. The guy had tattoos from head to toe and when I first saw him, I said to myself that he definitely didn't have what I wanted, not now, not ever. But as we learned from each other over the years, he really had a pretty good program and tremendous stories to tell. He was a good fellow and I started to like his heart. I had a lot of respect for him and even called him "friend." Well, I was walking up the steps to a new meeting hall; the old Pavilion had been torn down. This biker, named Frank, asked me where I had been. I told him that I had been so busy with business, politics and such, that I rarely had time for these meetings anymore. Frank promptly pulled me over to the side and reminded me that our primary purpose was to stay sober *and* help other alcoholics to achieve sobriety. "Yeah! So what?" I said. "So, you can't do that if you are not here," he said. "If you are not here, you are missing the whole boat. You are missing the point of recovery, of fellowship, of God." I walked in and sat down; I understood. Frank had given me the whole enchilada that night and what's more, I sat still and I knew it. The next day I went to a noon meeting, and there was Frank, asking me how my day was going. I told him, "Horrible!" and he replied, "You're not in the obituary column so it's not that bad." I smiled and thought, "Ugh! Go tell it to a newcomer." The not-so-funny thing was that the next day Frank's name was in the obituary column of the papers. He had been hit while riding on the Harley Davidson that he loved so much and died instantly.

It was a shock to everyone in the program and in the whole town, for that matter. Everybody knew and loved ole Frank. We had another sea burial and there were barely enough boats for all the people to make the trip. I didn't go out in my boat, but went with most of the fellowship on a big cruise boat that could hold about 300. People from all over the country started to show up on this beautiful day; it was truly incredible and inspiring. Where could you find a program that literally honored a guy who had scraped the bottom of the barrel; spent years in the big California prisons; turned his life around; and ended it with dignity, love and respect. Even the Hollywood movie stars whom Frank had sponsored came out that day to bid farewell. But those movie stars could never shine as high as Frank's star was shining and always would shine in the memory of this program. He influenced so many of us to pass on our primary purpose. I also benefitted from what Frank had to give and I am forever grateful to him. We put his ashes to rest and on the way back to the harbor, his last request was granted. Jimmi Hendrix was playing on the loudspeakers coming into port.

Frank was in his fifties when he died, with some twenty years of sobriety, but he didn't have much, materially. He was only a cashier at a restaurant, but I had never been to a funeral so big and seen so much love for anyone in my life. This truly was the result of living successfully. Here we were disembarking at the harbor and "Excuse Me While I Kiss The Sky" was playing on the sound system. It made you feel free, and I knew, for the first time, that it was important to die sober. It's not just important . . . *it's everything.*

My fifth year of sobriety was a milestone for me. I had been looking forward to it since I had gotten my chip for my first year. The night I was supposed to get my chip for five years, I was out on an overnight charter. There was some dope smoking going on and even though I didn't care for it, I didn't stop it. I knew where I should have been that night, but my priorities were backwards . . . I was out making money instead. Remembering Frank, I knew that I needed to change my thinking and was ready to do exactly that.

I went back to the group and picked up my five year chip. It was a very important event and I often share it with others. When I speak to a meeting, I tell how I turned my priorities around; about our primary purpose and Frank, the biker, who passed it on to me.

As time passed, I began to get more and more bored with business, life, pleasures on the islands and with my wife. It was coming up to 1991 and I was just about done with Hawaii. I had gone about as far as I could in my profession. It was time to move on. When you start to complain about every day being beautiful, the seasons that never change, the expenses, the wages, the people; it's time to leave. People still ask me how could I have left Hawaii. I always reply that I had reached the point that either I would settle there for the rest of my life or I was going to start searching for home. It was time for me to go home, and I felt it deep in my gut.

On August 10, 1991, a completely crazy man named Saddam Hussein invaded the small country of Kuwait. This event had a critical effect on the next year for me. Desert Storm, the world's answer to Hussein, was the beginning of the worst economic recession that I've felt in my lifetime. Suddenly the tourist business took a nosedive, leaving the Hawaiian economy reeling. Watching the war come to a quick and effective close, I lasted through a nonexistent winter season and went through whatever savings and financial cushion I had. When the economy never rebounded, my hopes for good business that summer were smashed. Business was upside down and I was barely hanging onto my house. Talk about a powerless feeling. I was supposedly doing everything right . . . making good business decisions, being frugal, standing by my principles, and being good at what I do. This insanity halfway around the world brought me to my knees in less than a year. We put the house and business up for sale and started to look at alternatives throughout the southern mainland; close to Louisiana, but not too close. In ninety days we had closed a deal on both my house and business. We luckily cashed out for a large profit and were on our way to the mainland. Throughout all this, Jamie and I were fighting about anything and everything possible. Did we want to just divvy up before we left for the mainland or try to hold it together and make things work? We decided to try to hold on and off we went . . . the Prodigal Son coming home after nine years of exile; looking forward to life in a place called home.

5 Coming Home

After watching some moderately valuable items go in the garage sale, we packed everything important in our lives into the moving van and started saying our goodbyes. For me the hardest thing was to say goodbye at the program and to leave the pets. We had tried very hard to find them good homes and hopefully we did. It hurt to let them go; they're like your children. Jamie was going to meet the family car, now onboard a ship bound for Oakland, and take the scenic route back to Louisiana. I flew into Lafayette, Louisiana, and stayed with my mom for a few weeks. While I was waiting for my wife to arrive, I tried to make a few bucks in some quick deals. Nothing seemed to fall into place; all I was doing was spinning my wheels. My mom had been divorced from my dad for about twelve years and by her own doing, had become a prisoner in her own home. She had regressed so bad that she and I couldn't get along at all. She argued even when you agreed with her. As a matter of fact, she was just as scared of me as she was when I had left, nine years ago. I was in my seventh year of sobriety, but while I was home, my mom walked around the house with a panic button to the alarm system strapped to her waist.

My father's reception was even colder. He told me to go back to my old oil field job. You must realize that my old job was as a surveyor out somewhere on an oil platform; sleeping in a sleeping bag; eating beanie wienies from a can and being at sea for weeks at a time. I was twenty-seven years old back then and forty-one years old now. There was no way I could ever go back to my old job even if I had wanted to. I just couldn't physically cut it anymore. After I was home a few days he even suggested that I go to work at Kmart. I would like to remind you that this man has a vast empire of wealth, employing hundreds of people and has countless connections. He could literally pick up the phone, call the governor of the state and within a few minutes be talking about last week's news. About this time I was beginning to feel the old resentments return again toward my family, my friends, and the whole town. Nothing had changed; everything was still the same in my little town, and I had been home only a week.

The only thing that had changed was me. I was sober and had choices to make. Did I want to buy this old garbage back? About seven A.M. one morning, my mom woke me, and as we had coffee, it was obvious she was troubled. I asked her why. She replied to my question saying she was worried about my sister who was in the middle of moving and it was very stressful, so she was worried for her. My chin must have been touching the floor as I turned red with anger. I looked at her dead in the face and said sharply, "My whole life is packed up on a freighter somewhere in the Pacific. My wife is on the road between California and the Grand Canyon. I have just sold off half of our belonging in a garage sale, lost our pets, not to mention that I don't have a job or a home for us to live in. Yet you are worried about my sister moving across town. Thanks, but no thanks, Ma. This kind of thing has the ability to make me at least crazy, if not drunk. I'm out of here." A friend named Brian and I took a trip to Gulf Shores, Alabama. I rented a beach home for a month so that my wife could get home and we could look for a place to live. At last some peace.

When Jamie arrived she took another week to spend with her family. She was on *vacation* while I was frantic trying to make the transition smoothly before the money ran out. I got a job as a deckhand on a snapper charter boat. What an ego buster . . .the king of blue marlin Kona fishing working on some snapper boat taking fish off the hooks for pilgrims who didn't know a snapper from a catfish. When a charter-booking desk became available, I leased the business only to find that life became a daily argument with twelve to fourteen charter boat captains. I was working ten to sixteen hours a day trying to get things straight and make this venture work. At some point I asked Jamie to take the duty at the store for a few hours so I could take a much needed break. Her reply was that she had more important things to do. The truth was that she was out buying antiques and furniture for a rented house like it was her only duty. She came home with a $1200 couch that I didn't even like and a brand new double bed for $700 that didn't have enough room in it for the two of us. Maybe that should have been a hint. Arguments over money started again, just like always, and I was beginning to have my worst year of sobriety ever.

It was Christmas time and we were almost out the door on our way to her mom's when she asked me for some money for the holidays. Apparently her unemployment check was late and she had some things on layaway. "Sure. How much do you need?" I said. "$1,400," she replied and I almost fell out of the chair.

She started to argue and in order to keep peace, I said sure I'll write the check. I looked up halfway through writing one thousand, four hundred, and she was still standing over me yelling. At that very moment I had what I now call a spiritual awakening. These days, when I see something from an entirely different angle and the result of it is a total change of my viewpoint, I define that as a spiritual awakening. While she was yelling and I was writing the check, I realized for the very first time that this person not only was incapable of loving me, but didn't even like me, and worst of all had no respect for me. I have had enemies in my life that didn't like me, but they damn sure respected me. I stared for a moment and tore up the check in front of her, commenting that I wouldn't be going to grandma's house for the holidays after all. She packed up her kid and off they went, leaving me alone and heartbroken for yet another Christmas season. I resolved right then and there that this was going to be the last Christmas I was going to spend by myself.

That Christmas day something happened that was to be the beginning of a new direction in my life. I went to a meeting that morning and was invited to a nice home with several people. We enjoyed good food and fellowship as we celebrated life on Christ's birthday. It was nice, but the best part came when I was saying "good-bye" and someone asked me if I would go to a prison meeting and share my story with the *cons* on that Christmas night. I immediately accepted and off we went. I stood before thirty or more prisoners and told my story; sharing my pain, strength, and hope. By the time I finished telling my story, I had been lifted out of myself and I felt gratitude for my sobriety as I had never felt before. The guys accepted me and asked if I would come back. I made a six-month commitment right then and there to be at the meeting every week.

The next week I heard someone else give their story and the week after that more of the same thing. On the third week, I was the only one there and the seats were arranged in an auditorium fashion. I started the meeting and asked if the only AA they had been exposed to was in the style of a speaker telling his story. The reply was, yes, and I had men break up the rows of chairs and put them in a circle. We started a new group of Alcoholics Anonymous that night, elected a chairperson and coffee maker, named the group, and started a series of discussion meetings for the next six months. If I gave anything at all to the prisoners that meant a great deal, it was starting the group. Later an incredible touching event happened to me and it is one of my favorite tales of sobriety. After one night's meeting, a giant of a man came to me and presented me with a question. I'll try to tell the story as well as I can remember so you will be able to picture the situation. He looked like a Mike Tyson-style prizefighter and walking up to me said, "Mr. Forrest, Mr. Forrest, do you think there are any answers in that Blue Book for me?" I said, "I don't know. Are you an alcoholic?" He said, "Oh, yes!" He had been locked up for crimes directly related to his drinking and drug use. He had been locked up for a little over a year. He didn't know if this had anything to do with his imprisonment, but someone had broken into his house a month ago and shotgunned his "ole lady" down and when the police got there, they found his two kids finger painting the walls with their mamma's blood. He immediately started to cry and blurted out that he couldn't stand himself anymore and couldn't even get any sleep at night. He wasn't sure if he hadn't paid one of the dope dealers off or if he had crossed someone bad enough to do that. He was angry, full of guilt and shame. Wiping his eyes dry, he looked at me for answers and only saw me standing there in astonishment. I told him that I wasn't a clinical psychiatrist or a professional counselor, not even a trained social worker. But if he was truly an alcoholic and would give himself completely to this simple program then, through the steps, we might be able to get him some comfort and some relief. Of course, I was willing to work with him any way I could. He thanked me and walked back to the cell block. I left to go back to the house in the pouring rain. I pulled over on the shoulder of the road before I got out of the prison grounds and just bawled, crying like a baby, saying, "Yes sir, God! I'll be privileged to do your work. I would be honored." I drove away and the message that Frank, "the Biker," had passed on to me kept burning into my mind . . . *"Our primary purpose is to stay sober and help other alcoholics achieve sobriety."* Not stay sober and make a million dollars. Not stay sober and get a new marriage. Not stay sober and start a new career. To stay sober and help another alcoholic achieve sobriety. It rang louder and louder in my head.

I never saw that particular inmate again. I don't know if he was transferred or just never came back to the meetings. He gave me something immeasurable that night that put me so much closer to God. I can't ever thank him enough. Later the state political system, allied itself with the meeting or group at that particular site, along with all the other institutions in the state. I stopped my participation with the prison group. I finished the commitment and went on to other things, but by far it had been some of the best work I had ever done, involving my sobriety thus far.

On another occasion I went on a mission into Mexico with a dentist to help the Mexican Indians with dental work. Everything had been planned for quite some time and I had looked forward to this event with a lot of enthusiasm as though it could turn out to be a life's calling for me. I have heard so much about the humanitarian spirit being the most rewarding. Nevertheless, as time approached for the day of departure, the dentist changed the roster of people in attendance. Originally there was to be himself, a longtime friend of his who had done this sort of thing before, and myself. Suddenly a 19-year-old female showed up and wanted to go along. She was brunette with long soft hair, makeup done in the most fashionable way, with a wardrobe consisting of short shorts climbing up the cheeks of her ass. Get the picture? Well, so did every Mexican for the next 3,000 miles.

Worst of all, the control of the entire project changed hands from the dentist being in charge of the trip, to whatever this little honey wanted to do and when and where she wanted to do it. She had the dentist agreeing with her on everything from political beliefs and strategies, to solving the world order. I couldn't take it anymore. I lost control of myself when she confronted me and I let her know exactly what I thought about her. Picture me two inches from her nose, screaming at volume twelve on the amplifier, right in her face. Had it been the old days I might of cut her throat and left her in a Mexican ditch. I certainly was that angry. These moments of insanity still persist. When I lose control of myself that badly, it is scary to admit what goes through my mind. I told the dentist to drop me off at the nearest airport. I had had enough. On the way home, I was dragging up all the depression that occurs every time I lose my temper and it was starting to hang off of me like sewage from the past. Six hundred dollars later I was home. Boy, was I fuming!

After the depression had hit, I found myself flying from Mexico City into Houston, and from Houston, I took a small hop into Lafayette, Louisiana. By the time I was on this commuter flight, it had become night and the best that I could do for myself was to stare aimlessly out the window, looking at the lights of cities below. With an occasional glance down at my shoes, I hoped to crawl into a deep hole somewhere. The seat next to me was empty and I prayed no one would take the available place. So far we had been up in the sky for awhile and that made it pretty safe that no one was going to sit next to me. I continued with this tormenting self-pity and looked up to see this airline captain asking if he could take the available seat. I said sure, with a disgusted look, hoping he would get the hint and not bother me. He didn't take the hint and started small talk by asking if I was going to Lafayette, and if I was from there. I answered yes to both questions and he said he was from Lafayette also, and that maybe he knew me. What was my name? "Forrest Dee," I told him. He said "Forrest Dee! I can't believe I'm sitting next to Forrest Dee. What are you doing with yourself these days, Forrest?" he asked. I told him that he had caught me at a bad time. I was just returning from a mission in Mexico that had failed and went on to say that I had been in recovery for some time and had put together eight years of sobriety. Most of my life these days was spent volunteering at the Salvation Army. I was now sharing a house with two other alcoholics and that we helped each other stay sober on a daily basis. One day, eventually I wanted to start a recovery house of my own and devote most of my life working with others, but today wasn't a good day and for the most part I felt like a failure.

He looked at me with a chin that was touching the floor, and said, "Failure! You're no failure, Forrest. Anyone who has a heart that big and can do all of that, is no failure. I guarantee you that. Look at me . . . All I do is drive an airplane. No, you're doing God's work and by no means is that ever a failure." At that very moment he had given to me what I had been searching for all of my life. He gave me dignity, pride, honor, trust, faith and hope. All of my life I had been looking for success and I was successful, in the purest way possible; God's way.

This is just another example of how things work totally opposite to how we perceive them to be. I wake up in the morning feeling strong and perceive that to be who I am, thinking that I could be governor of the state one day. If I wake up and feel weak, I feel like a failure in life. I have learned not to judge myself by emotions, because the emotions are not the truth. They are just how I feel in a fleeting moment.

Getting back to working with others, later during this time of my sobriety, I had been asked to help a friend I had known since I was nine years old. He had been in and out of jails since he was twenty-five years old. The guy had a bad combination of things that kept his life in a turmoil. His father kept funding him, and then when he got out of control, would have him locked up. His wife kept leaving him. His business was literally going up in a crack pipe and he was in jail for the umpteenth time. His family asked me, his wife asked me, and he asked me to help. I brought him to Alabama and we were roommates for over a year. He didn't have a chance in Lafayette, Louisiana. He was going to die there. He didn't have a driver's license so we were thrown together more than I would have liked. He did pretty well at first and even made a few prison meetings with me.

It had been very hard on both of us to live together and for me to try to sponsor him at the same time. I could foresee most of the problems and asked him to get someone else to sponsor him, but he always fell back into my lap. There were periods when he would hit on newcomers as they walked in the door of the program, knowing full well how much I despise that sort of action. He would go to bars and gambling halls till all hours of the morning. To this I have said, "Don't ever stand before a meeting and call me your sponsor because I don't want to be embarrassed by the results of your actions." He has not grown in his sobriety. He's just not drinking and using, and sometimes I have my doubts about that. He is selfish and self-seeking. Isn't that how we all got here? Why then do I expect anything different from him?

We were having a cookout one Sunday afternoon. I was putting two of the best steaks you have ever seen on the Bar-B-Q grill, when I turned to notice this extreme look of pain on his face. I continued what I was doing, and didn't mention anything right away. Going inside the house to mix up a little Bar-B-Q sauce and check on the baked potatoes, I passed him along the way. This look of intensity still persisted, and again I chose not to say anything. It wasn't long after I made the salads, retrieved the steaks, and buttered the potatoes, that we were sitting at the table, ready to eat. I looked up and there he sat with the same expression. Almost immediately I blurted out, "What's wrong with you, Man?" Using the biggest puppy dog eyes I had ever seen, he replied, "I'm just not happy with sobriety."

"O.K., so what do you want? Answer me that. What is it that you want?"

With the most intelligent look he could muster, he replied, "I just want to stay sober, do God's will, and help others."

"Oh balderdash!" I exclaimed. "You don't love God that much. Now what is it that you really want?"

A smile appeared and he asked, "Do you want the truth?"

"Of course, I want the truth. What do you want?"

He told me, "I want broads, I want a lot of money, and I want action. That's what I want."

I smiled and said, "Yeah, you want to go to jail too, because that's where you go every time you get any of that." I looked him straight in the eye and said, "O.K., let's try this again. Now clear your mind, take your time, take a deep breath," relax. He did everything I told him to do. "All right," I said, "Are you ready?"

"Yes," he said.

Again I asked him very calmly, "What is it that you want?"

He thought for a moment. He looked at me and thought some more. A look of astonishment came over him and he answered, "Man, I don't know. I don't know what I want. . . ."

We were both smiling by now, and I said, "Now ask me the same thing."

"O.K., Forrest, what do you want?"

"Man, I just want to enjoy this steak in front of me!" We burst out laughing so hard, we both almost fell out of our chairs.

I knew at that moment, I had grown up quite a bit in this program. I suddenly understood what one day at a time meant. I also understood that I had been miserable for a long time about what I wanted, or who had what, or who was doing what to whom. It was endless, and none of it really mattered much. All that did matter was enjoying the steak in front of me, and having a good day.

Later, this same person started to work in prison meetings and treatment centers and now has a pretty good head on his shoulders. He has become one of the biggest parts of my sobriety and has taught me more about myself and working with others than anyone else in my life. I am truly grateful to him, and love him dearly for it.

I continue to work with a good number of recovering people, no matter what the results. I volunteer at certain mental health facilities and this work has become the best days of my week. I have another roommate who is divorced and in the program. That makes three alcoholics living under one roof. It is a tough house, but it's lots of fun. I have learned quite a bit about myself from them. Just the other day I blew my stack with one of them and after the argument I looked at myself and saw how insignificant and insecure I am.

Now, trying to keep with the context of coming home, let me bring out that it really doesn't matter if I was rejected by my father when I got home, or that as soon as Jamie and I hit the mainland she deserted me and went to her mother's house. It really makes no difference if my mom has regressed to such a point of insecurity that she stays locked up in her house and can't be reached. The difficulty I am having with a career change in my own life is inconsequential. The desperation of loneliness that I feel can be dwarfed by the real problems of people so much less fortunate than I. The real meaning of life and all that is involved is not about me. I believe that for now, I am only a beam of light in God's Creation. A ray of sunshine in a moment of darkness. I am not important in the scheme of things and may or may not ever amount to much in this material world. It doesn't matter what kind of car I drive, what size house I live in or how much money I have in the bank. A hundred years from now, no one is ever going to remember who Forrest Dee was. What is important is that I can touch a child and change his life; reach out to a sick alcoholic and help keep him sober; or give to the ones less fortunate; then it has the power to change the world. I may not be here tomorrow, but if I have done something to help save a life then that ray of sunlight got through. That is what is important, powerful, immortal, and everlasting.

Due to all my experiences with drugs and alcohol I feel like I am qualified to go where God's angels cannot enter. Sometimes, with a foul mouth, I reach more people than professionals can. Most of the time when I do share from my heart, they can see the light shining through as real and genuine with truth and honor, giving them something to hope for. This truly does make us all who save lives and carry the torch, God's Prodigal Sons, and the prodigal sunlight in which a ray of hope may penetrate. Again I am not important, but the message of the light is and has been given to me by those before me, making it truly immortal. It is said that in order to keep it (God's Grace) you must give it away. In reality that is all I'm doing here. In the process of writing this book, I have been allowed to work all of the steps in the program over again. If it never goes any further than that, it has been worth every moment in front of this keyboard.

Some of the gifts I have received in sobriety which cannot be measured by any material methods are coming in leaps and bounds. I have waited patiently yet worked hard for them. Some of it didn't turn out the way I expected, some turned out better than I thought, but all in all, evidence proves that the promises do come to us. My dad invited me to his eightieth birthday party. I spent the entire weekend with family and never a cross word was spoken. The real miracle that weekend was my daughter, Courtney, spent the weekend with us too. Yes! The tiny baby who I was to never see again was growing up and now able to meet her real dad. She is fourteen years old and even though it's true that I missed her childhood, I hope not to miss her life.

My sister and I have a relationship that we both didn't think possible. She and I have been able to share some of our innermost thoughts with one another. Most of all, I know she would be there for me if I needed her.

I have friends galore. Once I was having a bad week and I must have gotten calls from six different states, from people who love and care for me, the most support and love coming from people in the program.

My God and I are very close today. I have strong faith for the most part of my life. Though sometimes I will falter, I feel His Love through it all. I don't ask for selfish things. Well, every 1,000 days or so I do, and it is given to me whether it is good for me or not. Once I asked and I waited, asked and waited again. I took a moment and very seriously talked to God, and He gave me my wish, as if to say, *Yes, I'm here with you and do hear you.* I know He does hear and feels every bit of my pain because I offer it to Him. Soon the days start to come with comfort and start adding up. Quickly the healing of the body, mind, and finally, spirit, arrive. Insanity leaves only to return when called on from old behavior and lower human instincts, then soon leaves again due to awareness of what we have conjured up. Life becomes good . . . Love is allowed to flow and hearts are warmed. For this there is no measuring cup. All I know is my cup runneth over and I am truly blessed. For me there is no death, only a reuniting with my Holy Father.

You know, I do not think that this world is of God's world. The injustice I see in today's news is extremely demoralizing. The Exxon

Valdez, Chernobyl, and the condition of the ozone are good examples of the total destruction of our environment. Deterioration of the animals and nature are becoming alarmingly evident, with no end in sight. Beyond the pollution and economic ruin we are having to bear, do not let us forget the exhausting religious wars that have gone on for centuries and the prejudice and hatred that are still lethal to us all. Sometimes it overwhelms me and I think, "What's the use? Nobody is listening and worse still, nobody cares." I feel the need to stand up for what is good, right, and just in the world today. Again a speck of light shining through the darkness and despair; giving birth to lasting ideas of promise, a new freedom and a new happiness. Laying down principles that are morally right, guiding us to some honest joy. The seed is planted, the garden grows, and the tides change in favor of a movement called something like Save The Planet, Save Our Lives, Save Our Children. I believe that Alcoholics Anonymous will be the guiding force in which some of the world relief from this despair will come. It has been evident what can happen when two drunks get together in the spirit of helping each other. Now it is time for that positive force to be used in the practical world. Maybe it is time to not be so anonymous. Maybe it is time to share these principles in all of our affairs.

If we have learned anything in the height of our disease, surely we know that we have affected all who have been around us. That being the case, we should know that we can also affect all those whose lives we touch in sobriety. We would then help make the world in which we live a more positive place to be. *Remember Attraction Rather Than Promotion.* Stand up and be counted for upholding these principles as a guide to the way things are supposed to be in God's paradise before we let money, material things, and instant gratification take over and control us all. We can at the very least return to that paradise ourselves and be examples for our future leaders, *our children!*

Today I am in God's Kingdom. Today I do care. Today the sun shines through my eyes and into my heart. Today God lives in my soul and I am whole. Today I just didn't pick up a drink and let the rest of it happen. I love you all in Alcoholics Anonymous as God's children and fellow humans carrying the torch of light for those who are still suffering. Today as the sun sets here on the Gulf of Mexico, I watch it go down to the very last ray, setting off a green flash not seen very often in this area. As it disappears over the horizon, I realize that now and forever, I have truly come home. God Bless you all!

Through much of the commitment and devotion in helping the alcoholic who is still suffering, it has become obvious that the prisoners being released from jail, the crack addicts completing treatment, and the alcoholics left out on the street, do not have a chance of recovery when returned to their old environment. Therefore, I am relinquishing all funds and donations from the proceeds of this book to the foundation which has been set up for the construction and acquisition of three-quarter homes for the members in recovery.

If you have enjoyed meeting me and agree with the direction my life has taken, please help us provide clean and sober environments to those less fortunate than most. Give generously to the Clearview Foundation so that others may have a chance to follow the ray of sunlight to happy destiny.

If you would like an autographed
copy of *One Way Home,*
a $12 donation or a contribution may be sent to:

The Clearview Recovery Foundation
101 Nicole Place
Daphne, AL 36526

The Author

Born in Lafayette, Louisiana. Attended University of Southwestern Louisiana. Received U.S.C.G. Captain's License in Hawaii. Owner / Operator of Chiripa Sportfishing, Inc., out of Honolulu, Hawaii. Now residing in Daphne, Alabama. Member of the Board of Directors of the Eastern Shore Recovery Foundation. Chairman of the Clearview Recovery Foundation. Has done extensive volunteer work in prisons and mental health facilities throughout the Mobile area.